ABOUT THIS WORKBOOK

The Introduction – Overview, The Chart, Additional Information, and Diagrams – In these sections, you will find the essential information about a specific part of the reference table. These areas contain much information, so read each section slowly and carefully to achieve full comprehension of the material.

SET 1 – Questions and Answers – Set 1 group of questions will test your understanding of a specific section of the reference table. It is highly recommended that you first read and have a good knowledge of the introduction pages. Once you have mastered this, the correct answer to each question will be apparent. Try all questions in Set 1, and then correct your work by going to the Answers for Set 1, which are located at the end of the section. The explanation should be clear enough to help you understand any mistakes you have made. If not, ask your teacher for more assistance.

SET 2 – Questions – The answers to these questions are provided in a separate answer key. It's "Show Time"; time to prove to yourself and to your teacher that you know the information for this part of the reference chart. You and your teacher will interact to see how well you have done in this area.

All of us at Topical Review Book Company hope that by mastering the Earth Science Reference Tables, your understanding of Earth Science will be more complete and your grades will improve.

The author:
William Docekal
Retired Earth Science Teacher

4th Edition
© 2019, Topical Review Book Company, Inc. All rights reserved.
P. O. Box 328
Onsted, MI 49265-0328
www.topicalrbc.com

Physical Setting/Earth Science
Reference Table Workbook
Table of Contents

Physical Setting/Earth Science Reference Table Workbook
Table of Contents

Radioactive Decay Data

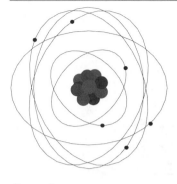

RADIOACTIVE ISOTOPE	DISINTEGRATION	HALF-LIFE (years)
Carbon-14	$^{14}C \longrightarrow {}^{14}N$	5.7×10^3
Potassium-40	$^{40}K \begin{smallmatrix} \nearrow {}^{40}Ar \\ \searrow {}^{40}Ca \end{smallmatrix}$	1.3×10^9
Uranium-238	$^{238}U \longrightarrow {}^{206}Pb$	4.5×10^9
Rubidium-87	$^{87}Rb \longrightarrow {}^{87}Sr$	4.9×10^{10}

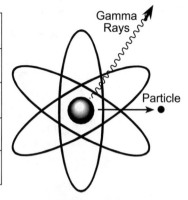

Gamma Rays

Particle

Overview:

Radioactive elements release nuclear particles called alpha and beta particles and energy in the form of gamma rays. When these particles are emitted from the nucleus of a radioactive atom, the element changes, becoming a different element. This is called radioactive disintegration. Each radioactive element has its own unique half-life (the time it takes for half of the radioactive element to disintegrate), which is constant and cannot be changed by environmental influences such as heat and pressure. Using the half-life, scientists have developed methods to date objects containing a radioactive element. This radioactive dating technique is our basis for the geologic timeline.

The Chart:

The Radioactive Decay Data chart shows 4 radioactive isotopes. Over time, radioactive elements will decay and change into different non-radioactive stable element(s). In the Disintegration column, it shows the radioactive element's symbol and the non-radioactive element's symbol(s) into which it would eventually decay or disintegrate. As shown, radioactive carbon-14 will change into non-radioactive nitrogen-14. The time it takes for half of a radioactive element to decay into its decay product is called its half-life. For example, if we start with a 100% radioactive C-14 sample, after the 1st half-life, 50% is still radioactive and 50% is non-radioactive, and 5,700 (5.7×10^3) years have passed. After the 2nd half-life, 25% is C-14 and 75% is N-14, and 11,400 years have passed (5,700 yr + 5,700 yr). This process continues until the radioactive element has completely changed into the non-radioactive or decayed substance.

Additional Information:

- In a radioactive isotope, radiation gets weaker in time, but the half-life stays the same.

- Carbon-14 is found in organic substances, such as bones, wood, and shells and has a relatively short half-life compared to the other 3 given radioactive isotopes in the above chart.

- Carbon-14 is used to date relatively young organic substances, less than 100,000 years old. Thus dinosaur bones, being millions of years in age, are too old to be dated using C-14.

- Uraniuim-238 has a half-life of 4.5×10^9 years, which equals 4.5 billion years (10^9 = billion).

- Uraniuim-238 is used to date very old (billions of years) inorganic substances, such as rocks and meteorites. This radioactive element was used to arrive at the estimated time of origin of Earth.

- On a graph, the radioactive isotope decreases from 100% moving downward toward 0%, while the non-radioactive element increases from 0%, moving upward toward 100%.

Diagrams:

1. **Radioactive Decay** – The models below represent the decay of radioactive atoms to stable atoms after their first and second half-lives.

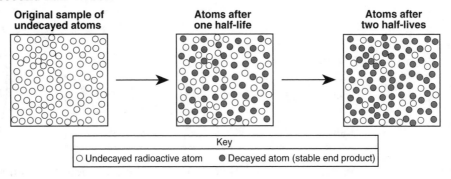

Original sample of undecayed atoms → Atoms after one half-life → Atoms after two half-lives

Key
○ Undecayed radioactive atom   ● Decayed atom (stable end product)

2. **Half-Life Graph** – This is a typical decay graph of any radioactive isotope. As shown by the Time axis, this radioactive isotope has a half-life of 5,000 years. After each half-life, the radioactive material is reduced by 50%, but the half-life does not change. The shaded pie graph is another way to illustrate the half-life decay process.

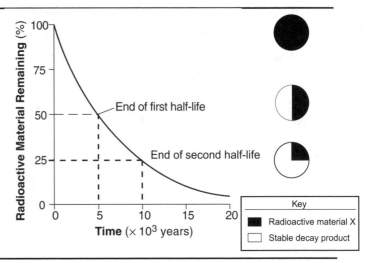

End of first half-life

End of second half-life

Key
■ Radioactive material X
□ Stable decay product

3. **Half-Life Flowchart** – The flowchart shows the percentage ratio of the radioactive isotope carbon-14 to its decay product, non-radioactive nitrogen-14, for the first two half-lives. The Time chart gives the time in years after two half-lives of ^{14}C.

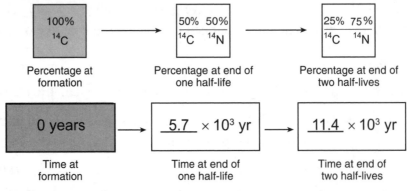

| 100% ^{14}C | → | 50% 50% ^{14}C ^{14}N | → | 25% 75% ^{14}C ^{14}N |

Percentage at formation | Percentage at end of one half-life | Percentage at end of two half-lives

| 0 years | → | 5.7 × 10³ yr | → | 11.4 × 10³ yr |

Time at formation | Time at end of one half-life | Time at end of two half-lives

4. **Radioactive Dating** – The wood, being geologically relatively young, was dated using radioactive carbon-14. The dinosaur skull, being approximately 110 million years old, would have no measurable ^{14}C. To date this skull, a radioactive isotope with a much longer half-life was used.

wood
28,000 yr old

dinosaur skull
110 million yr old

1. Radioactive carbon-14 dating has determined that a fossil is 5.7×10^3 years old. What is the total amount of the original C^{14} still present in the fossil?

   (1) 0%   (3) 50%
   (2) 25%   (4) 75%   1 _____

2. A sample of wood found in an ancient tomb contains 25% of its original carbon-14. The age of this wood sample is approximately

   (1) 2,800 years   (3) 11,400 years
   (2) 5,700 years   (4) 17,100 years   2 _____

3. How much of an 800-gram sample of potassium-40 will remain after 3.9×10^9 years of radioactive decay?

   (1) 50 grams   (3) 200 grams
   (2) 100 grams   (4) 400 grams   3 _____

4. A graph of the radioactive decay of carbon-14 is shown to the right.

   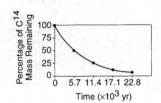

   Which graph correctly shows the accumulation of nitrogen-14, the decay product of carbon-14, over the same period?

   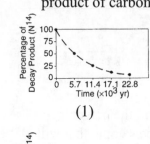

   (1)

   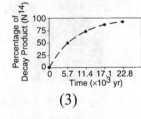

   (3)

   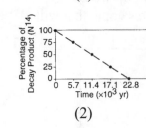

   (2)

   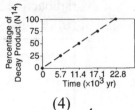

   (4)   4 _____

5. The absolute age of a rock is the approximate number of years ago that the rock formed. The absolute age of an igneous rock can best be determined by

   (1) comparing the amounts of decayed and undecayed radioactive isotopes in the rock
   (2) comparing the sizes of the crystals found in the upper and lower parts of the rock
   (3) examining the rock's relative position in a rock outcrop
   (4) examining the environment in which the rock is found   5 _____

6. An archaeologist found an ancient skeleton estimated to be 10,000 to 25,000 years old. Which radioactive isotope would be most useful for finding the age of the skeleton?

   (1) carbon-14   (3) uranium-238
   (2) potassium-40   (4) rubidium-87   6 _____

7. Which graph best represents the radioactive decay of uranium-238 into lead-206?

   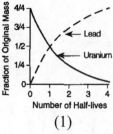

   (1)

   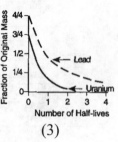

   (3)

   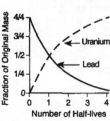

   (2)

   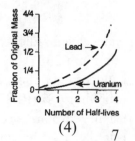

   (4)   7 _____

8. The table below gives information about the radioactive decay of carbon-14. [Part of the table has been left blank for student use.]

Half-Life	Mass of Original C-14 Remaining (grams)	Number of Years
0	1	0
1	$\frac{1}{2}$	5,700
2	$\frac{1}{4}$	11,400
3	$\frac{1}{8}$	17,100
4		
5		
6		

What is the amount of the original carbon-14 remaining after 34,200 years?

(1) $\frac{1}{8}$ (3) $\frac{1}{32}$

(2) $\frac{1}{16}$ (4) $\frac{1}{64}$ 8 _____

9. The diagram below represents the radioactive decay of uranium-238. Shaded areas on the diagram represent the amount of

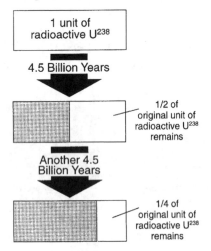

1 unit of radioactive U²³⁸

4.5 Billion Years

1/2 of original unit of radioactive U²³⁸ remains

Another 4.5 Billion Years

1/4 of original unit of radioactive U²³⁸ remains

(1) undecayed radioactive uranium-238 (U238)
(2) undecayed radioactive rubidium-87 (Rb87)
(3) stable carbon-14 (C14)
(4) stable lead-206 (Pb206)   9 _____

10. Base your answer from the diagram below, which represents a model of a radioactive sample with a half-life of 5,000 years. The white boxes represent undecayed radioactive material and the shaded boxes represent the decayed material after the first half-life.

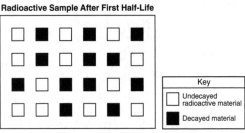

Radioactive Sample After First Half-Life

Key
☐ Undecayed radioactive material
■ Decayed material

How many *more* boxes should be shaded to represent the additional decayed material formed during the second half-life?

(1) 12   (3) 3
(2) 6   (4) 0   10 _____

11. Which process could be indicated by the expression below?

$$U^{238} \rightarrow Pb^{206}$$

(1) crystallization of minerals in basalt
(2) chemical weathering of marble
(3) radioactive decay in granite
(4) ozone depletion in the atmosphere   11 _____

12. Base your answer from the diagram below, which represents a sample of a radioactive isotope.

Sample before decay

Key
■ Radioactive isotope
☐ Decay product

In the box below shade in the percentage of the radioactive isotope sample that will remain after 2 half-lives.

13. A sample of wood that originally contained 100 grams of carbon-14 now contains only 25 grams of carbon-14. Approximately how many years ago was this sample part of a living tree?

    (1) 2,850 yr   (3) 11,400 yr
    (2) 5,700 yr   (4) 17,100 yr   13 ____

14. The characteristic of the radioactive isotope uranium-238 that makes this isotope useful for accurately dating the age of a rock is the isotope's
    (1) organic origin
    (2) resistance to weathering and erosion
    (3) common occurrence in sediments
    (4) constant half-life   14 ____

15. The accompanying diagram represents the present number of decayed and undecayed atoms in a sample that was originally 100% radioactive material. If the half-life of the radioactive material is 1,000 years, what is the age of the sample represented by the diagram?

    (1) 1,000 yr   (2) 2,000 yr   (3) 3,000 yr   (4) 4,000 yr   15 _____

Base your answers to question 16 from the accompanying graph. The graph represents the decay of radioactive material **X** into a stable decay product.

16. *a)* What is the approximate half-life of radioactive material **X**?

    (1) 5,000 yr   (3) 50,000 yr
    (2) 10,000 yr   (4) 100,000 yr   a _____

    b) Which graph best represents the relative percentages of radioactive material **X** and its stable decay product after 15,000 years?

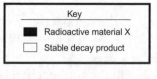

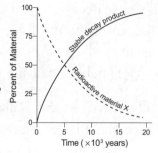

    b _____

    c) Each of the objects below has different amounts remaining of the original radioactive material **X**. Which object is most likely the oldest?

| (1) Coal 10% of the radioactive material remains | (2) Shell 41% of the radioactive material remains | (3) Wood 33% of the radioactive material remains | (4) Bone 52% of the radioactive material remains |

    c _____

17. A student filled a graduated cylinder with 1,000 milliliters of water to represent a radioactive substance. After 30 seconds, the student poured out one-half of the water in the cylinder to represent the decay occurring within the first half-life. The student repeated the process every 30 seconds. How much water was left in the cylinder at the 2-minute mark?

(1) 12.5 mL   (2) 62.5 mL   (3) 125.0 mL   (4) 250.0 mL   17 _____

18. Radioactive C^{14} was used to determine the geologic age of old wood preserved in a glacier. The amount of C^{14} in the old wood is $\frac{1}{8}$ the normal amount of C^{14} currently found in the wood of living trees. What is the geologic age of the old wood?   _____ years

19. State *one* difference between dating objects with the radioactive isotope carbon-14 (C^{14}) and dating objects with the radioactive isotope uranium-238 (U^{238}).

_____

Base your answers to question 20 on the accompanying diagram, which represents a model of a radioactive decay of a particular element. The diagram shows the decay of a radioactive element (☐) into the stable decay element (■) after one half-life period.

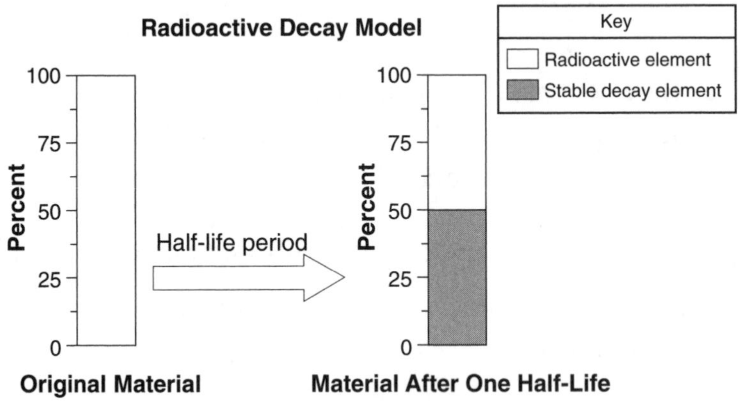

Radioactive Decay Model

Key
☐ Radioactive element
■ Stable decay element

Original Material   Material After One Half-Life

20. *a)*   If the radioactive element in this model is carbon-14, how much time will have passed after one half-life?

_____ years

b)   If the radioactive element in this model is uranium-238, how much time will have passed after one half-life? _____ years

c)   On the Radioactive Decay Model shown to the right, shade in the amount of stable decay element present after the second half-life period.

21. State the name of the radioactive isotope that has a half-life that is approximately the same age as the estimated time of the origin of Earth. _____

22. Identify the element formed when rubidium (^{87}Rb) undergoes radioactive decay. _____

23. As a radioactive element decays, what happens to its half-life period? _____

1. **3** The Radioactive Decay Data chart shows that the half-life of C-14 is 5.7×10^3 yr which equals 5,700 years. This is the same age as the fossil. When a radioactive element has undergone radioactive decay of one half-life, 50% of the original radioactive substance remains, while the other 50% has changed to a different element.

2. **3** After the first half-life, 50% of the original radioactive substance remains. At the second half-life, 25% of the original radioactive substance remains. The half-life of C-14 is given as 5.7×10^3 yr or 5,700 years. Even though the original radioactive material decreases over time, the half-life of that element does not change. Thus, for the second half-life of C-14, the total time that passed would be 11,400 years (5,700 + 5,700).

3. **2** The half-life of potassium-40 (K-40) is 1.3×10^9. After one half-life, or 1.3×10^9 yr, K-40 would have decayed by 50% and 400 g would be radioactive. After 2 half-lives, or 2.6×10^9 yr, 200 g would still be radioactive K-40. After the 3 half-lives, or 3.9×10^9 yr, only 100 g of K-40 would be radioactive.

4. **3** The Radioactive Decay Data chart shows that C-14 decays into N-14. During this decay process, the total amount of C-14 and N-14 must equal 100%. Thus, as the C-14 undergoes radioactive decay and becomes less, the amount of N-14 will increase. At the first half-life, the percentage of these two elements will be 50%-50%. After two half-lives, the percentage will be: C-14, 25%, and N-14, 75%. This process continues until C-14 is 0% and N-14 is 100%.

5. **1** Scientists use radioactive dating to arrive at the absolute ages of rocks. Within rocks are radioactive isotopes that undergo a constant decay, changing from the original radioactive element to a different element. The time it takes for half of the original element to change into its decay element is given the term half-life. The absolute age can be obtained by comparing the ratio of the radioactive substance to its decay element and using the known half-life.

6. **1** Carbon-14 is used to date relatively young organic substances, especially bones. The Radioactive Decay chart shows that the half-life of C-14 is 5,700 years. This half-life is very short compared to the other given radioactive elements. C-14 is used to date relatively young organic remains that are less than 100,000 years old.

7. **1** The Radioactive Decay Data chart shows that uranium-238 will disintegrate into the decay product lead-206. As the uranium disintegrates into lead, the total amounts of the two elements must equal 100%. After the first half-life, both elements will be 50%-50%. After the second half-life, the remaining U-238 will be 25%, and the decay product Pb-206 will be 75%. The radioactive material continues to decrease toward 0%, while the decay product increases toward 100%.

8. 4 As C-14 undergoes radioactive decay, the amount of C-14 will be cut in half as the years increase by 5,700 for each half-life. Following this procedure, 1/16 = 22,800 years, 1/32 = 28,500 years, and 1/64 of the original C-14 remains after 34,200 years.

9. 4 The Radioactive Decay Data chart shows that uranium (U-238) undergoes radioactive decay changing to lead (Pb-206). For each half-life, the amount of U-238 decreases by 50%, while the non-radioactive element Pb-206 increases by 50%. The white area must represent the decaying radioactive uranium, while the gray area represents the increasing decay product of lead, Pb-206.

10. 2 After one half-life, half of the radioactive substance will decay into a new substance. The original radioactive sample had 24 squares. After the first half-life 12 squares would still be radioactive (white squares), while 12 squares would have decayed (the black squares). This is shown in the diagram. The second half-life would cut the 12 remaining radioactive squares by half, leaving 6 white radioactive squares, while shading 6 more boxes to make a total of 18 shaded (decayed material) squares.

11. 3 The Radioactive Decay Data chart shows that radioactive uranium disintegrates, or decays, into the stable element lead, Pb-206. Uranium is found in certain minerals contained in the igneous rock granite.

12.

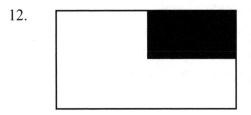

Explanation: After the first half-life, the ratio of radioactive to non-radioactive elements is 50% to 50%. After the second half-life, the ratio of radioactive (black area) to non-radioactive (white area) elements is 25% to 75%.

Remember:

Carbon 14 is used to date relatively young organic (once living) material.

Specific Heats of Common Materials

MATERIAL	SPECIFIC HEAT (Joules/gram • °C)
Liquid water	4.18
Solid water (ice)	2.11
Water vapor	2.00
Dry air	1.01
Basalt	0.84
Granite	0.79
Iron	0.45
Copper	0.38
Lead	0.13

Heats up Faster

Heats up Faster

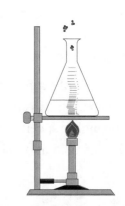

Overview:

Substances do not all heat up or cool down at the same rate. When heat (measured in joules) is absorbed by substances that do not go through a phase change, the temperature increases. Measuring how fast substances increase in temperature compared to water gives us this chart, called Specific Heats of Common Materials.

The Chart:

The Specific Heat chart shows how fast a substance heats up and cools down compared to liquid water. The higher the specific heat value is, the slower that substance will undergo a temperature change compared to a lower specific heat value. Water has the highest specific heat value, 4.18 Joules/gram•°C, of all common substances. That is why water needs to absorb a lot of heat before it gets hot. As shown by the chart, if the masses were equal, lead, with the lowest value 0.13, would heat up, and cool down the fastest of all the given substances. Copper having a value of 0.38 would heat up slower than lead but faster than all substances having a higher specific heat value. Granite and basalt, both being rocks, are used to represent land materials. From their respective specific heat values, one can see that land heats up much faster than water (about 5x's as fast).

Additional Information:

- If a substance heats up fast, it also cools down fast. Land not only heats up faster than water, but it also cools down faster than water. This is why the sand on a beach becomes much hotter than the body of water during a summer day, and why sand cools down faster than the water at night.

- Large bodies of water have a major affect on the climate of coastal areas. Water, having a large specific heat, heats up slowly and cools down slowly. This property of water causes coastal land areas to have a smaller annual temperature range and less extreme temperatures compared to inland cities at the same latitude.

- Heat is transferred by three methods: by conduction in solids, by convection in fluids (gases and liquids) and by radiation – as electromagnetic energy.

Diagram:

This diagram shows conduction in metal. In conduction heat is transferred by molecular vibrations through solid materials. Metals are excellent conductors of heat.

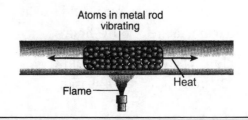

Atoms in metal rod vibrating

Flame

Heat

Heat

1. Liquid water can store more heat energy than an equal amount of any other naturally occurring substance because liquid water

   (1) covers 71% of Earth's surface
   (2) has its greatest density at 4°C
   (3) has the higher specific heat
   (4) can be changed into a solid or a gas   1 _____

2. How do the rates of warming and cooling of land surfaces compare to the rates of warming and cooling of ocean surfaces?

   (1) Land surfaces warm faster and cool more slowly.
   (2) Land surfaces warm more slowly and cool faster.
   (3) Land surfaces warm faster and cool faster.
   (4) Land surfaces warm more slowly and cool more slowly.
                                                              2 _____

3. The same amount of heat energy is added to equal masses of lead, iron, basalt, and water at room temperature. Assuming no phase change takes place, which substance will have the *smallest* change in temperature?

   (1) lead   (3) basalt
   (2) iron   (4) water   3 _____

4. Five-gram samples of granite, basalt, iron, and copper at room temperature are placed in a beaker of boiling water. Which sample would reach a temperature of 60°C first?

   (1) copper   (3) granite
   (2) iron   (4) basalt   4 _____

5. Equal masses of basalt, granite, iron, and lead received the same amount of solar energy during the day. At night, which of these materials cooled down at the fastest rate?

   (1) basalt   (3) iron
   (2) granite   (4) lead   5 _____

6. Large oceans moderate the climatic temperatures of surrounding coastal land areas because the temperature of ocean water changes

   (1) rapidly, due to water's low specific heat
   (2) rapidly, due to water's high specific heat
   (3) slowly, due to water's low specific heat
   (4) slowly, due to water's high specific heat
                                                              6 _____

7. Compared to the climate conditions of dry inland locations, the climate conditions of locations influenced by a nearby ocean generally result in

   (1) hotter summers and colder winters, with a larger annual range of temperatures
   (2) hotter summers and colder winters, with a smaller annual range of temperatures
   (3) cooler summers and warmer winters, with a larger annual range of temperatures
   (4) cooler summers and warmer winters, with a smaller annual range of temperatures   7 _____

8. Land surfaces of Earth heat more rapidly than water surfaces because

    (1) more energy from the Sun falls on land than on water
    (2) land has a lower specific heat than water
    (3) sunlight penetrates to greater depths in land than in water
    (4) less of Earth's surface is covered by land than by water   8 _____

9. Pieces of lead, copper, iron, and granite, each having a mass of 1 kilogram and a temperature of 100°C, were removed from a container of boiling water and allowed to cool under identical conditions. Which piece most likely cooled to room temperature first?

    (1) copper   (3) iron
    (2) lead   (4) granite   9 _____

10. Which group of substances is arranged in order of decreasing specific heat values?

    (1) iron, granite, basalt
    (2) copper, lead, iron
    (3) dry air, water vapor, ice
    (4) liquid water, ice, water vapor   10 _____

11. Equal volumes of the four samples shown below were placed outside and heated by energy from the Sun's rays for 30 minutes.

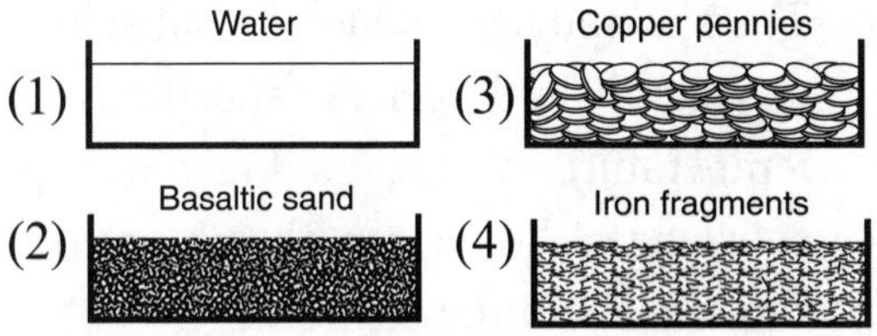

    The surface temperature of which sample increased at the slowest rate?

    (1) water
    (2) copper pennies
    (3) basaltic sand
    (4) iron fragments   11 _____

12. After sunset, what can be expected with the cooling rate of soil compared to the cooling rate of water?

    (1) The soil will cool faster because it is a good reflector.
    (2) The soil will cool faster because it has a lower specific heat.
    (3) The water will cool faster because it is a good absorber.
    (4) The water will cool faster because it has a higher specific heat.   12 _____

13. Equal masses of copper, lead and basalt were placed in direct sunlight for the same time interval. Assuming they were all at the same initial temperature, place the name of the substance in the correct position below, showing how they would be expected to increase in temperature.

    A 10°C increase – _____   A 6°C increase – _____   A 4°C increase – _____

14. During the winter, explain why locations near the Atlantic Ocean have air temperatures that are warmer than locations farther inland, at the same latitude.

_____

1. 3   The specific heat value of a substance shows how fast an object heats up or cools down compared to water. The larger the specific heat value of a substance, the more heat it takes to increase its temperature. Water, having the largest specific heat value, heats up the slowest of common substances. Water can absorb and store more heat compared to other substances undergoing the same temperature change.

2. 3   The greater the specific heat value of a substance, the slower that substance heats up and cools down compared to other substances. Land is represented by the two given rocks in the chart, basalt and granite. Both have a lower specific heat value compared to water. This shows that land heats up faster and cools down faster than water. This is evident at a beach when, during the day, the sand gets hot quickly and cools down quickly after sunset. The water temperature, however, does not change much day to day.

3. 4   As shown by the specific heat values, water would heat up and cool down the slowest, experiencing the smallest change in temperature. Lead, having the lowest specific heat, would undergo the greatest change in temperature.

4. 1   All of the given substances have the same mass and are at the same initial temperature. Copper, having the lowest specific heat value, would be the first to reach 60°C.

5. 4   Because all conditions are the same, the substance with the lowest specific heat value will cool down the fastest. From the Specific Heat chart, lead has the lowest specific heat value of the given choices.

6. 4   Water has the highest specific heat value, therefore oceans heat up slowly and cool down slowly. Because of this, large bodies of water tend to modify the yearly temperature range of the adjacent land. Large bodies of water are a major climate factor for coastal land areas.

7. 4   Water, having a high specific heat, will heat up and cool down more slowly than land surfaces. Thus, during the winter, the oceans will be warmer than the land, and during the summer, the oceans will be cooler than the land surfaces. This causes the ocean to have a modifying affect on the climate of coastal cities. These factors produce a smaller annual temperature range for coastal cities compared to inland cities.

   Note: For answers 6 and 7, see second bullet in Additional Information, page 9.

Overview:

When ice absorbs heat (measured in joules), its temperature increases until it reaches 0°C. At this temperature, it undergoes the phase change of melting. Now, as in all phase changes, the temperature remains the same

Heat energy gained during melting	334 J/g
Heat energy released during freezing	334 J/g
Heat energy gained during vaporization	2260 J/g
Heat energy released during condensation	2260 J/g
Density at 3.98°C	1.0 g/mL

until the phase change is completed. When the ice has completely melted and heat is still being applied, the temperature of the water will increase as the water absorbs heat energy. At 100°C (at sea level), the water has reached its boiling point and begins to boil. It will stay at this temperature during this phase change while continuing to absorb heat. During the phase changes of condensation and freezing, heat is released. In fact, within a hurricane a tremendous amount of heat is released when water vapor condenses forming its clouds. This is a major source of energy for the hurricane.

The Chart:

For ice to melt (s → ℓ), it must absorb heat. The chart shows that for every gram of ice that melts 334 joules are absorbed or gained. For every gram of water that freezes (ℓ → s), it must release or lose 334 joules. Notice that in these two phase changes, the number of joules are the same, but the difference is whether energy is being absorbed or being released. Both of these phase changes occur at 0°C, and the temperature remains the same while the phase change is occurring.

In vaporization (boiling or evaporation), a liquid changes into a gas (ℓ → g). For vaporization of one gram of water, the chart shows that 2,260 joules are absorbed. When condensation occurs (g → ℓ), each gram of water vapor releases 2,260 joules as it changes back to liquid water. Notice how much more heat is absorbed to vaporize one gram of water compared to the amount of heat absorbed to melt one gram of ice.

The density value of water is 1.0 g/mL, and as shown on the chart this occurs at 3.98°C (which can be rounded up to 4°C). When water warms up from this temperature, its density slightly decreases. This is why warm water rises, being displaced upward by the cooler, sinking, denser water.

Additional Information:

- Evaporation absorbs heat; therefore, the process of evaporation is a cooling one. The evaporation of a liquid off a surface will cause that surface to cool down.

Diagrams:

1. **Melting and Freezing** – Ice absorbs 334 joules/gram as it melts. When water changes back to ice, 334 joules/gram are released.

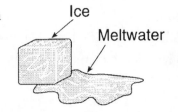

Temperature	
Ice	0°C
Meltwater	0°C
Air	7°C

2. **Vaporization** – In evaporation, water is being vaporized. In this phase change, water needs to absorb much energy, 2,260 joules/gram. The Sun adds tremendous amounts of energy to the atmosphere as it evaporates tons and tons of water each day.

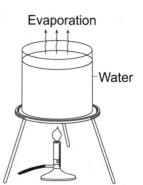

3. **Condensation** – As shown on the diagram, condensation is forming as moist air hits the cold glass. When this occurs, the water vapor releases 2,260 joules/gram as it condenses. Remember, clouds are formed as water vapor condenses, producing very small floating water droplets while releasing heat.

Set 1 — Properties of Water

1. Which phase change requires water to gain 2260 joules per gram?

   (1) solid ice melting
   (2) liquid water freezing
   (3) liquid water vaporizing
   (4) water vapor condensing   1 _____

2. During which phase change of water is the most energy released into the environment?

   (1) water freezing
   (2) ice melting
   (3) water evaporating
   (4) water vapor condensing   2 _____

3. Which statement best explains why water in a glass becomes colder when ice cubes are added?

   (1) The water changes into ice.
   (2) Heat flows from the water to the ice cubes.
   (3) Ice has a higher specific heat than water.
   (4) Water is less dense than ice.   3 _____

4. How many joules are required to evaporate 1 gram of boiling water?
   (1) 1 J   (3) 334 J
   (2) 80 J   (4) 2,260 J   4 _____

5. What is the total number of joules required to melt 100 grams of ice at 0°C to liquid water at 0°C?

   (1) 5,400 J   (3) 33,400 J
   (2) 8,000 J   (4) 226,000 J   5 _____

6. When water changes to ice, its density decreased because
   (1) the mass of the water decreased as it changed to ice
   (2) the mass of the ice decreased when it was formed
   (3) the water volume expanded as it changed to ice
   (4) the water volume decreased as it changes to ice   6 _____

7. How many joules (J) of heat energy are released by each gram of water vapor that condenses to form cloud droplets?

   7_____ J

8. Which process requires water to gain 334 joules of energy per gram?

   (1) vaporization　(3) melting
   (2) condensation　(4) freezing　　8 _____

9. Which energy statement is correct when 10 grams of water vapor condenses?

   (1) 2,260 J is released
   (2) 22,600 J is gained
   (3) 22,600 J is released
   (4) 3,340 J is released　　9 _____

10. Which process requires the most absorption of energy by water?

    (1) melting 1 gram of ice
    (2) condensing 1 gram of water vapor
    (3) vaporizing 1 gram of liquid water
    (4) freezing 1 gram of liquid water

    10 _____

11. During which phase change does water release the most heat energy?

    (1) freezing　(3) condensation
    (2) melting　(4) vaporization　　11 _____

12. Which diagram correctly shows the processes that change the states of matter?

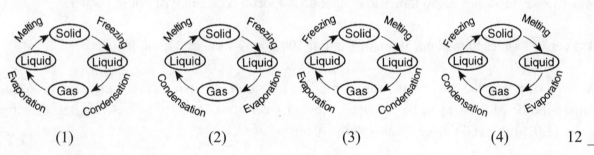

(1)　　　　　　(2)　　　　　　(3)　　　　　　(4)　　　12 _____

13. When water at 90°C is cooled down to 4°C its density would slightly

    (1) increase　　　　(2) decrease　　　　(3) remains the same　　　13 _____

14. Give the two phase changes that release energy.

    _____ and _____

15. Give the two phase changes that gain energy.

    _____ and _____

16. How many joules of heat energy are required to evaporate 2 grams of water from a lake surface?

    _____ J

Answers

Set 1

1. 3   The Properties of Water chart shows that during vaporization, 2,260 J/g are needed. When the boiling water absorbs this amount of energy, one gram of liquid water will change intowater vapor (steam).

2. 4   When ice melts and when water evaporates (vaporizes) heat is absorbed or gained. During freezing and condensation heat is released. This is shown on the Properties of Water chart. During freezing, each gram of water releases 334 joules of heat, and during condensation, 2,260 joules are released.

3. 2   When ice is added to water, the ice will absorb heat from the water lowering the water's temperature. As shown on the Properties of Water chart, each gram of ice will gain or absorb 334 joules as it melts.

4. 4   Water needs to absorb 2,260 joules to evaporate (vaporize) one gram of boiling water.

5. 3   It takes 334 joules to melt one gram of ice. For 100 grams, it takes 33,400 joules.

6. 3   As water freezes the mass stays the same, but the volume expands. This increase in volume, causes the density of ice to be .9 g/cm^3. Now, the density of the ice is less than the density of water (1.0 g/mL), and the ice floats. (Note: g/cm^3 = g/mL)

7.   Answer: 2,260 J

    Explanation: When condensation occurs, the water vapor condenses to water droplets. This process releases 2,260 J/g.

Remember:

Absorbs heat = melting and vaporization

Releases heat = freezing and condensation

Metric Ruler

Overview:

The metric ruler is not included on the reference tables, but you will be using a metric ruler throughout the year in Earth Science, so it is included in this workbook.

The Ruler:

On the side of this page is a metric ruler. The numbers on the ruler represent centimeters (cm). The smaller lines represent millimeters (mm). There are 10 mm for each cm. This scale is used to measure given objects, especially when measuring sediment sizes that are at actual scale. It is also extensively used with the eccentricity equation.

For length measurements, you must know the following:

One centimeter is equal to 10 mm.
One meter is equal to 100 cm.
One meter is equal to 1,000 mm.
One kilometer is equal to 1,000 m.

Additional Information:

- Conversions   Decimal point move
  km ↔ m   3 places
  m ↔ cm   2 places
  cm ↔ mm   1 place

- When converting a larger unit to a smaller unit, the decimal moves to the right.
  When converting a smaller unit to a larger unit, the decimal moves to the left.

- One inch = 2.54 cm, One meter = 39.4 inches, One kilometer = 0.62 miles

Examples:

a) 8.2 cm = 82 mm

b) 62 cm = 620 mm

c) 67 mm = 6.7 cm

d) 4.5 mm = .45 cm

e) 33 cm = .33 m

f) 125 cm = 1.25 m

g) .50 m = 50 cm

h) 3.25 m = 325 cm

i) 660 m = .660 km

j) 2.2 km = 2,200 m

> Questions dealing with the **Metric Ruler** will be found throughout this workbook.

Eccentricity

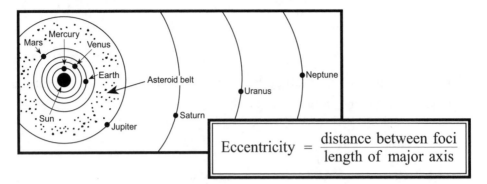

$$\text{Eccentricity} = \frac{\text{distance between foci}}{\text{length of major axis}}$$

Overview:

When early astronomers accepted that the Sun was the center of our solar system (heliocentric theory), they assumed that planets, comets, and other celestial objects revolved around the Sun in circular orbits. In time, it was proven that they do not revolve in circular orbits, but in elliptical orbits. These orbits are not perfectly round, but slightly flattened, giving them an oval shape. It is easy to describe a circular orbit, but how does one describe an elliptical orbit? This was solved by mathematicians using a formula called eccentricity. Think of this term as a measurement of how much the shape of an ellipse deviates from a circle. Eccentricity (e) is a value that shows how elliptical an orbit is. All ellipses have eccentricities (e) lying between zero and one ($0 < e < 1$).

The Equation:

To arrive at the eccentricity value for a given orbit, first measure the distance (d) between the two foci (F_1 and F_2). Located at one of these foci may be a star (like our Sun), while the other focus is at an imaginary position in space. Next measure the length of the major axis (l). Dividing, $\frac{d}{l}$ gives the eccentricity or e value. This e value has no units. The larger the eccentricity, the more elliptical the orbit will be. Turn to the Solar System Data chart (see page 198) and locate the Eccentricity of Orbit column. Mercury, having the largest eccentricity or e value, must have the most elliptical orbit of all planets. Venus, having the smallest eccentricity, has almost a circular orbit. The Earth's orbit, having a very low eccentricity of 0.017, would appear to be almost circular, but because e is more than 0, Earth's orbit is slightly elliptical.

Diagram:

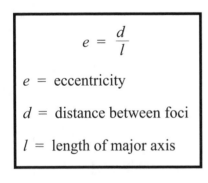

$$e = \frac{d}{l}$$

e = eccentricity

d = distance between foci

l = length of major axis

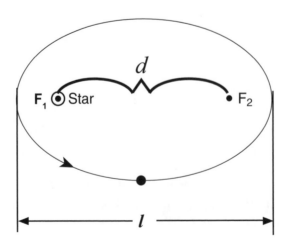

Example:
Calculate the eccentricity of this elliptical orbit.

Solution:

$$e = \frac{d}{l}$$

$$e = \frac{4.7 \text{ cm}}{7.0 \text{ cm}}$$

$$e = .67$$

Additional Information:

- When drawing an ellipse, if the foci distance is increased, the eccentricity increases and the ellipse takes on a "flatter" appearance .

- Due to the elliptical shape of orbits, when the orbiting bodies (planets, comets, etc.) revolve around the Sun, their distance from the Sun changes, at times being closer and at times being farther away.

- When an orbiting object is closer to the Sun (star), it speeds up. This is because the gravitational attraction with the Sun increases, causing the increase in orbital speed.

- When an orbiting object is farther from the Sun, the gravitational attraction with the Sun decreases, causing a decrease of orbital speed.

- Earth is closer to the Sun in winter (its perihelion), thus having its greatest orbital speed during this time.

- Earth is farther from the Sun in summer (its aphelion), thus having its slowest orbital speed during this time.

Set 1 — Eccentricity

Note: For some questions, you will need to use the Solar System Data chart on page 198.

1. Which object is located at one foci of the elliptical orbit of Mars?

   (1) the Sun   (3) Earth
   (2) *Betelgeuse*   (4) Jupiter   1 _____

2. The diagram below shows the elliptical orbit of a planet revolving around a star. The star and F_2 are the foci of this ellipse. What is the approximate eccentricity of this ellipse?

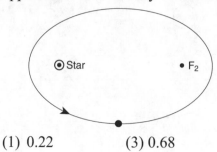

   (1) 0.22   (3) 0.68
   (2) 0.47   (4) 1.47   2 _____

3. Which planet has the most eccentric orbit?

   (1) Mercury   (3) Neptune
   (2) Venus   (4) Saturn   3 _____

4. Which of the following eccentricity values would produce the most eccentric orbit?

   (1) .009   (3) .01
   (2) .09   (4) .34   4 _____

5. Since Earth has an elliptical orbit, the

   (1) distance between the Sun and Earth varies
   (2) distance between the Sun and the other focus varies
   (3) length of Earth's major axis varies
   (4) length of Earth's period of revolution varies   5 _____

6. Which planet has an orbit with an eccentricity most similar to the eccentricity of the Moon's orbit around Earth?

   (1) Earth   (3) Uranus
   (2) Jupiter   (4) Saturn   6 _____

Base your answers to question 7 on the diagram below, which represents the elliptical orbit of a planet traveling around a star. Points A, B, C, and D are four positions of this planet in its orbit.

7. *a)* The calculated eccentricity of this orbit is approximately

   (1) 0.1   (3) 0.3
   (2) 0.2   (4) 0.4   a _____

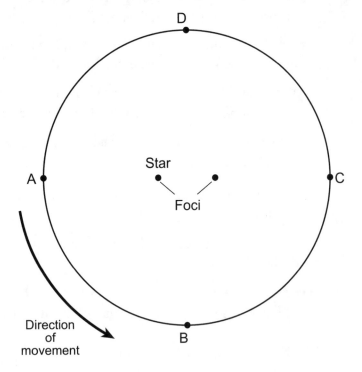

   b) The gravitational attraction between the star and the planet will be greatest at position

   (1) *A*   (3) *C*
   (2) *B*   (4) *D*   b _____

   c) What planet could this orbit represent?

   _____

Base your answers to question 8 on the diagram below, which represents an exaggerated model of Earth's orbital shape. Earth is closest to the Sun at one time of year (perihelion) and farthest from the Sun at another time of year (aphelion).

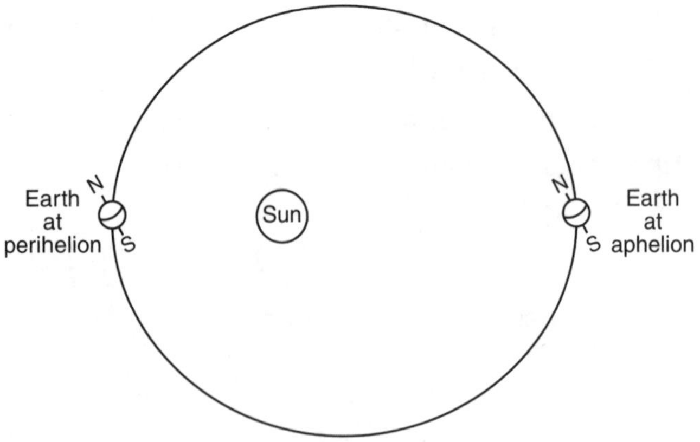

8. *a)* State the actual geometric shape of Earth's orbit.

   _____

   b) The Sun is at F_1. Within the given diagram, place a small **X** where F_2 would be located.

   c) Give the season when Earth is at perihelion. _____

   d) Give the season when Earth is at aphelion. _____

   e) What season is Earth's orbital velocity the slowest? _____

Base your answers to question 9 on the below diagram of the ellipse.

9. *a*) Write out the eccentricity equation.

b) From the given ellipse, substitute the correct values into the equation and calculate the eccentricity of the ellipse.

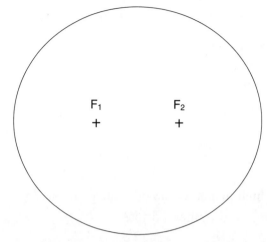

F₁   F₂
+   +

c) State how the eccentricity of the given ellipse compares to the eccentricity of the orbit of Mars.

_____

d) Describe how the shape of the orbit would change, if the distance between F_1 and F_2 increases.

_____

Base your answers to question 10 on the accompanying diagram which shows a model of the orbital path of Earth and the partial orbital path of Jupiter around the Sun. A partial orbit of another celestial object, labeled object *A*, is also shown. Celestial object *A* is a natural object that is part of our solar system.

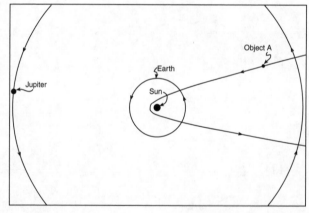

10. *a*) Identify object *A*. _____

b) Give a statement on the eccentricity of object *A* compared to the eccentricity of Earth's orbit.

(Object size not drawn to scale)

_____

c) What is causing the orbital velocity of object *A* to be increasing?_____

d) Why is the comet considered to be part of our solar system?

_____

11. Which planet has the least elliptical orbit?

    (1) Jupiter
    (2) Mars
    (3) Venus
    (4) Saturn   11 _____

12. Which planet has an orbit shape most similar to Jupiter's orbit?

    (1) Mercury   (3) Saturn
    (2) Uranus   (4) Mars   12 _____

13. Which diagram shows a planet with the *least* eccentric orbit?

(Key: • = planet ✳ = star)

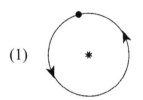

(1)

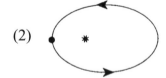

(2)

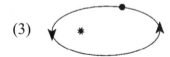

(3)

(4)   13 _____

14. The actual orbits of the planets are

    (1) elliptical, with Earth at one of the foci
    (2) elliptical, with the Sun at one of the foci
    (3) circular, with Earth at the center
    (4) circular, with the Sun at the center   14 _____

15. The diagram below is a constructed ellipse. F_1 and F_2 are the foci of the ellipse. The eccentricity of this constructed ellipse is closest to the eccentricity of the orbit of which planet?

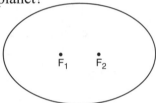

    (1) Mercury   (3) Saturn
    (2) Earth   (4) Venus   15 _____

16. Earth is farthest from the Sun during the Northern Hemisphere's summer, and Earth is closest to the Sun during the Northern Hemisphere's winter. During which season in the Northern Hemisphere is Earth's orbital velocity greatest?

    (1) winter   (3) summer
    (2) spring   (4) fall   16 _____

17. One of the foci for the Moon's orbit would be the

    (1) Sun   (3) Mars
    (2) Earth   (4) Venus   17 _____

18. The accompanying diagram represents the elliptical orbit of a spacecraft around the Sun. Calculate the eccentricity of the spacecraft's orbit following the directions below:

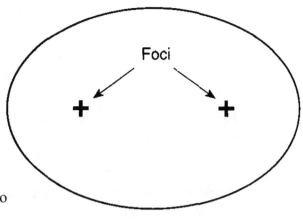

a) Write the equation for eccentricity.

b) Substitute measurements of the diagram into the equation and calculate the eccentricity and record your answer in decimal form.

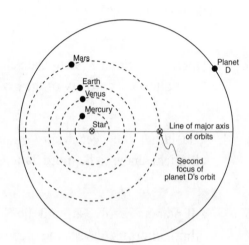

19. Describe the eccentricity of planet D's orbit relative to the eccentricities of the orbits of the planets shown in our solar system.

_____

_____

20. A student constructed the accompanying elliptical orbit of a Moon revolving around a planet. The foci of this orbit are the points labeled F_1 and F_2.

Describe how the shape of the elliptical orbit would change if the distance between the foci points was 2.5 cm.

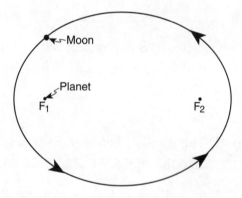

_____

21. State the relationship between Earth's distance from the Sun and Earth's orbital velocity.

_____

1. 1   All planets in our solar system revolve around the Sun in elliptical orbits. For all planets, the Sun is located at one foci, the other is a point in space.

2. 3   Using a metric ruler, measure the distance between the foci (Star and F_2) and the length of the major axis. Substitute into the eccentricity equation.

       Solution: $e = d/l$, $e = 3.4$ cm/5.0 cm $= 0.68$ Remember, no units are assigned to the answer.

3. 1   Open to the Solar System Data chart (see page 198) and locate the Eccentricity column. Mercury has the highest eccentricity of the given choices.

4. 4   The larger the eccentricity value is, the more eccentric the orbit will be. This will cause the elliptical shape to be more flattened.

5. 1   All planets revolve around the Sun in elliptical orbits with the Sun being at one foci (F_1). Having elliptical orbits, causes the planets to be closer to the Sun and farther from the Sun at some time in their orbit around the Sun.

6. 4   Open to the Solar System Data chart. In the Eccentricity of Orbit column, the Moon's value is given as 0.055. Saturn's value is 0.054. Both of these orbits have very similar elliptical shapes, but not the same size.

7. *a)* 2   The distance between the foci is 1.6 cm. The distance of the major axis is 7.8 cm. Substituting into the equation and solving, $e = 1.6$ cm $/7.8$ cm $= 0.2$.

   b) 1   The closer a planet is to a star, the stronger the gravitational attraction will be. This causes the planet to increase in its orbital speed.

   c)   Answer: Mercury

       Explanation: Mercury has the closest eccentricity (.206) to the eccentricity answer of 0.2 (see Solar System Data chart, page 198).

8. *a)* Answer: Elliptical *or* Slightly elliptical

    Explanation: All planets revolve around the Sun in elliptical orbits.

b)

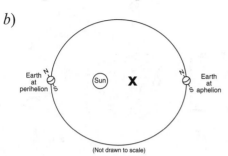

(Not drawn to scale)

Explanation: Credit is awarded if **X** is positioned to the right of the Sun as shown.

c) Answer: Perihelion – Winter   Explanation: During winter, Earth is closer to the Sun, and the northern axis tilts away from the Sun.

d) Answer: Aphelion – Summer   Explanation: During summer, Earth is farther from the Sun, and the northern axis tilts toward the Sun.

e) Answer: Summer   Explanation: During summer, Earth is farther from the Sun, and it is experiencing its slowest orbital velocity.

9. *a)* Eccentricity $= \dfrac{\text{distance between foci}}{\text{length of major axis}}$

b) $e = \dfrac{2.3 \text{ cm}}{6.8 \text{ cm}} = .33 \ or \ .34$ Explanation: Measure the distance between the foci, F_1 to F_2 and measure the length of the major axis.

c) Answer: The given ellipse has a higher eccentricity than the orbit of Mars.

       or   The eccentricity of Mars' orbit is less than that of the given ellipse.

       or   The orbit of Mars is more circular than the given ellipse.

Explanation: The larger the eccentricity value is, the more elliptical the orbit will be. The eccentricity of the given ellipse is .33. Mars' eccentricity is 0.093, which is less than the given elliptical orbit. Thus, Mars' orbit is less elliptical or rounder compared to the given ellipse.

d) Answer: becomes more elliptical

Explanation: Increasing the distance of the foci causes the shape to become more elliptical.

10. *a)* Answer: Comet

Explanation: Comets are under the gravitational attraction of the Sun. This causes them to revolve around the Sun in very elliptical orbits.

b) Answer: The comet's eccentricity is a higher value than that of the Earth's.

       or The Earth's eccentricity is a lower value than that of the comet.

Explanation: The greater the eccentricity value is, the more elliptical the orbit will be.

c) Answer: Gravity *or* gravitational attraction to the Sun

Explanation: The gravitational attraction to the Sun becomes stronger as the comet gets closer. In response to this increase of gravitational force, the comet speeds up.

d) Answer: See explanation 10*a*.

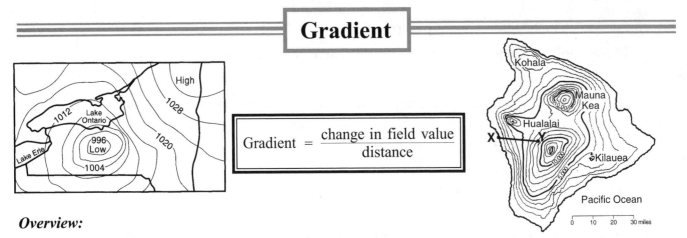

Gradient

$$\text{Gradient} = \frac{\text{change in field value}}{\text{distance}}$$

Overview:

A field is an area in which measurements can be obtained at any point within the field region. Some examples of fields are: an elevation field, as shown on a topographic map containing contour lines; an air pressure field measured in millibars (mb), that is displayed on a weather map using isobars; or a temperature field connected by isotherms. The gradient (G) equation gives the average change within the field from two given points or positions. The gradient may indicate the steepness of a land surface, or it could show the strength (magnitude) of an air pressure field that could indicate the potential of a severe weather situation.

The Equation:

The change in field value is the difference between two values. These values may be given to you, or you might have to determine their values from two given points on some type of field map with drawn isolines. Use the isoline interval to obtain the values of the two given points within the field. For the distance between the two points, use the distance scale usually located at the base of the field map. Divide to get the answer. No credit will be awarded if the proper units are not included.

Example: The topographic map shows a stream crossing several contour lines and passing through points X and Y. Elevations are measured in feet. What is the approximate gradient between point X and point Y?

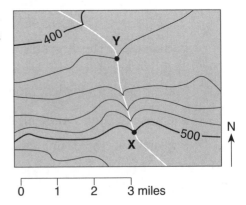

Solution: $G = \dfrac{500 \text{ ft} - 420 \text{ ft}}{2 \text{ mi}} = \dfrac{80 \text{ ft}}{2 \text{ mi}} = 40 \text{ ft/mi}$

Explanation: From the map, the contour interval (spacing) is 20 ft. The change in field value from Y to X would be 80 ft. Using the given distance scale, the distance is 2 miles.

Additional Information:

- When isolines become closer, the gradient is increasing.

- Contour lines that are closely spaced on a topographic map indicate a steep slope (a hill, or mountain, etc.).

- When isobars on a weather map are tightly spaced, the pressure gradient is strong, and that area will be experiencing windy conditions.

- When contour lines cross a stream they form a "V" that always points upstream. This can be seen in the Example diagram.

- A topographic profile is a cross-sectional (side) view of the relief of the terrain between two locations.

1. Which equation can be used to correctly calculate the air-pressure gradient between two locations?

   (1) gradient = $\dfrac{\text{change in air pressure (mb)}}{\text{average air temperature (°F)}}$

   (2) gradient = $\dfrac{\text{change in air pressure (mb)}}{\text{distance (km)}}$

   (3) gradient = $\dfrac{\text{change in distance (km)}}{\text{air pressure interval (mb)}}$

   (4) gradient = $\dfrac{\text{change in air pressure (mb)}}{\text{air pressure interval (mb)}}$

   1 _____

2. A topographic map shows two locations, X and Y, one half mile apart. From the contour lines, the elevation of X is 800 feet and Y is 750 feet. What is the gradient between the two locations?

   (1) 12.5 ft/mi (3) 50 ft/mi
   (2) 25 ft/mi   (4) 100 ft/mi   2 _____

3. Point A and point B are locations 0.24 mile apart on a ski slope in northern New York. Point A has an elevation of 1,560 feet and point B has an elevation of 1,800 feet. What is the gradient between these points?

   (1) 60 ft/mi   (3) 500 ft/mi
   (2) 240 ft/mi   (4) 1,000 ft/mi   3 _____

4. On each topographic map below, the straight-line distance from point A to point B is 5 kilometers. Which topographic map shows the steepest gradient between A and B?

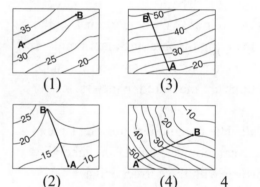

   (1)   (3)

   (2)   (4)   4 _____

Base your answer to question 5 on the accompanying map showing ocean depths, measured in meters, off the coast of Massachusetts. Points A, B, and C represent locations on the ocean floor.

5. Calculate the average ocean-floor gradient between point A and point B. Label your answer with the correct units.

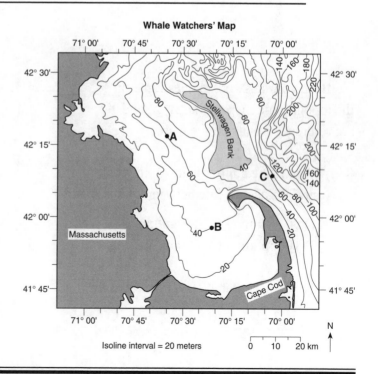

Whale Watchers' Map

Isoline interval = 20 meters   0 10 20 km

Base your answers to question 6 on the topographic map. Elevations are in feet. Points *A* and *B* are locations on the map.

6. *a*) What is the gradient along the straight line between points *A* and *B*?

    (1) 10 ft/mi   (3) 25 ft/mi

    (2) 20 ft/mi   (4) 35 ft/mi   a _____

   b) Which direction is Green River flowing?

    (1) south   (3) northeast

    (2) east   (4) northwest   b _____

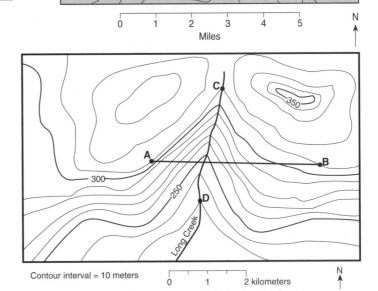

Base your answer to question 7 on the accompanying topographic map. Points *A*, *B*, *C*, and *D* are reference points on the map. Elevations are measured in meters.

7. Calculate the gradient of Long Creek between points *C* and *D* and label the answer with the correct units.

Contour interval = 10 meters

Base your answers to question 8 on the topographic map. Points *A*, *X*, and *Y* are reference points on the map.

8. *a*) What is the approximate gradient along the straight dashed line between points *X* and *Y*?

    (1) 50 m/km   (3) 150 m/km

    (2) 100 m/km   (4) 300 m/km   a _____

   b) Give an acceptable elevation value for *A*.

_____ m

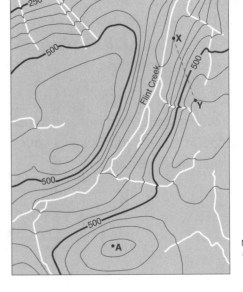

Contour interval = 50 meters

Base your answer to question 9 on the topographic map below. Elevations are expressed in feet.

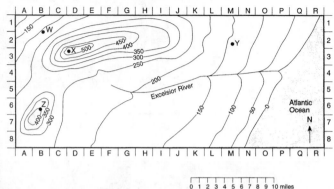

9. What is the gradient of the entire length of the Excelsior River?

(1) 0.1 ft/mi   (2) 11 ft/mi   (3) 24 ft/mi   (4) 48 ft/mi   9 _____

Base your answers to question 10 on the accompanying map. The map shows a portion of the eastern United States with New York State shaded. The isolines on the map indicate the average yearly total snowfall, in inches, recorded over a 20-year period.

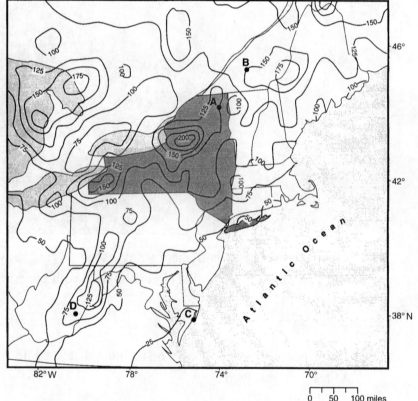

10. *a)* The average yearly snowfall for position *D* is closest to

(1) 75 in   (3) 100 in
(2) 85 in   (4) 125 in

b) From the given latitude and longitude readings, give the coordinates of the area that has the greatest average yearly total snowfall.

_____ ° N, _____ ° W

c) What is the approximate average yearly total snowfall gradient between locations *A* and *B*?

(1) 0.25 in/mi   (2) 2.50 in/mi   (3) 0.40 in/mi   (4) 4.00 in/mi   c _____

11. How can you tell a steep gradient on a topographic map?

12. The accompanying topographic map shows locations *X* and *Y*.

What is the approximate gradient between *X* and *Y*?

(1) 15 ft/mi
(2) 20 ft/mi
(3) 30 ft/mi
(4) 60 ft/mi

12 _____

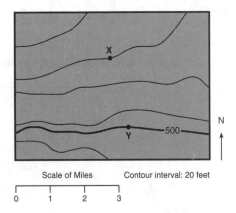

Scale of Miles   Contour interval: 20 feet

0   1   2   3

Base your answers to question 13 on the weather map. The isobars show air pressures, in millibars. Points *A* and *B* indicate locations on the map.

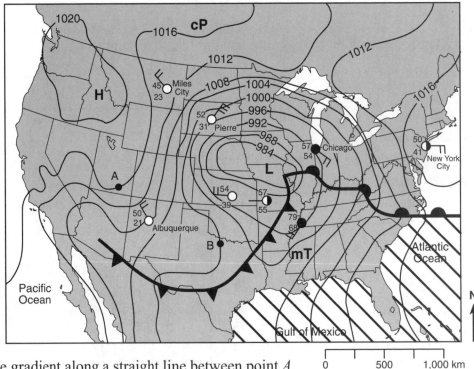

13. *a)* Calculate the pressure gradient along a straight line between point *A* and point *B* on the map. Label your answer with the correct units.

b) Describe the evidence shown on the map that indicate strong winds are blowing between Miles City and Pierre.

_____

c) Give an acceptable pressure reading for Miles City. _____ mb

d) How does one know that the pressure gradient is greater between Chicago and New York City compared to point *A* and point *B*?

_____

_____

1. 2   The gradient equation is $G = \dfrac{\text{change in field value}}{\text{distance}}$.

    Air pressure is measured by a barometer in units of mb. The change in field value will be the difference in air pressure from two pressure readings. The distance will be in kilometers.

2. 4 $\text{Gradient} = \dfrac{\text{change in field value}}{\text{distance}}$ $G = \dfrac{800 \text{ ft} - 750 \text{ ft}}{.5 \text{ mi}} = \dfrac{50 \text{ ft}}{.5 \text{ mi}} = 100 \text{ ft/mi}$

3. 4   The change in elevation is the difference between the two elevations, which is 240 feet (1800 ft – 1560 ft). The distance is 0.24 mile.

    Solution: $G = \dfrac{1800 \text{ ft} - 1560 \text{ ft}}{0.24 \text{ mi}}$ $G = 240 \text{ ft}/0.24 \text{ mi} = 1000 \text{ ft/mi}$

4. 4.   The distances from A to B are the same for each map, but the change in field values for map 4 shows the greatest change in elevation, 35 m (50 m – 15 m). This distance would have the steepest gradient of the given maps. Also in map 4, A-B crosses more contour lines.

5.   Answer: $G = .5$ m/km

    Explanation: From the map and the given contour interval, point A is 60 meters and point B is 40 meters. The difference of these two elevations gives the change in field value. To obtain the distance from A to B use the given distance scale.

    Solution: $G = \dfrac{60 \text{ m} - 40 \text{ m}}{40 \text{ km}}$ $G = 20 \text{ m}/40 \text{ km} = .5$ m/km

6. a) 3   The change in field value is, 400 ft – 300 ft = 100 ft. The distance is 4 miles.

      Solution: $G = \dfrac{400 \text{ ft} - 300 \text{ ft}}{4 \text{ mi}}$ $G = 100 \text{ ft}/4 \text{ mi} = 25 \text{ ft/mi}$

   b) 4   Rivers flow from high elevations to low elevations. Also the "V" that the contour lines form when they cross over a stream always bends upstream.

7.   Answer: $G = 28.6$ m/km (± 3.0 m/km)

    Explanation: The contour interval is 10 meters. Using this interval, the elevation of point C is 310 meters. Point D is two lines below the 250 meter isoline, making the elevation of point D 230 meters. Using the given scale, the distance between points C and D is close to 2.8 km.

    Solution: $G = \dfrac{310 \text{ m} - 230 \text{ m}}{2.8 \text{ km}}$ $G = 80 \text{ m}/2.8 \text{ km} = 28.6$ m/km

8. a) 3 The contour interval is 50 m. From the 500 m contour line, the elevation decreases toward X and increases toward Y. The elevation of Y is 600 m and X is 300 m.
      Solution: $G = 600 \text{ m} - 300 \text{ m}/2 \text{ km} = 300 \text{ m}/2 \text{ km} = 150$ m/km.

   b) Answer: Any value from 651 m to 699 m

    Explanation: Position A is above the 650 m contour line, but its elevation must be below 700 m because no 700 m contour line is given.

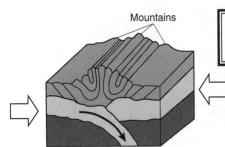

Mountains

$$\text{Rate of change} = \frac{\text{change in value}}{\text{time}}$$

Overview:

Certain earth science events occur quickly, like an earthquake or a landslide. Other events may take thousands or even millions of years, like mountain building or the erosion of a landscape region. In most cases, earth scientists will calculate the rate of change for an event. From these rates of changes, comparisons and conclusions can be made about the event.

The Equation:

All events take time, and the event may have caused a change to the original values (measurements). When this situation occurs, a rate of change can be obtained by using the above equation.

Example 1: If a pile of sand 10.0 feet high is eroded over 60 days to a height of 7.0 feet, what is the rate of change (erosion) for this pile of sand?

Solution: Rate of change (Rc) $= \dfrac{10.0 \text{ ft} - 7.0 \text{ ft}}{60 \text{ d}}$ $Rc = \dfrac{3.0 \text{ ft}}{60 \text{ d}} = 0.05 \text{ ft/d}$

Explanation: The change in value involves subtracting the two variables, in this case 10.0 ft – 7.0 ft, giving a change in value of 3.0 ft. Next, as the equation shows, divide by the time.

Example 2: Calculate the rate of change in the average monthly temperature for Omaha during the two-month period between October and December, as shown on the graph.

Solution: Rc = 16°C/2 mo = 8°C/mo

Explanation: The change in value is obtained from the graph. In October the temperature was 12°C and in December the temperature is –4°C. This gives a change in temperature value of 16°C in 2 months. For one month the Rc = 8°C/mo.

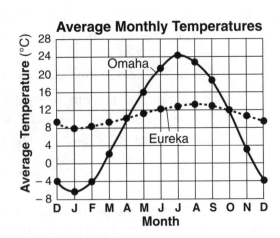

Additional Information:

* The rate of movement of plate tectonics has become very accurate by using laser and GPS measuring devices.

1. The temperature of water in a container was 60°C. Ten minutes later, the water temperature was 35°C. What was the rate of cooling of the water?

   (1) 25°C/min
   (2) 2.5°C/min
   (3) 35°C/min
   (4) 3.5°C/min   1_____

2. The graph below shows temperature readings for a day in April. The average rate of temperature change, in Fahrenheit degrees per hour, between 6 a.m. and noon was

   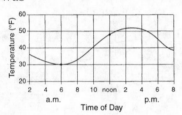

   (1) 6°F/hr   (3) 3°F/hr
   (2) 8°F/hr   (4) 18°F/hr   2_____

3. The rate of temperature change for the water in cup *A* for the first 10 minutes was approximately

	Temperature of Water (°C)	
Minute	Cup *A*	Cup *B*
0	90	20
1	88	20
2	86	20
3	85	21
4	83	21
5	82	22
6	81	22
7	80	22
8	79	22
9	78	23
10	77	23

   (1) 0.77°C/min
   (2) 1.3°C/min
   (3) 7.7°C/min
   (4) 13.0°C/min   3_____

4. Calculate the average daily rate of movement of the hurricane from August 24 to August 28. Follow the directions given below.

   a) Write the equation used to determine the rate of change.

   b) Substitute data into the equation.

   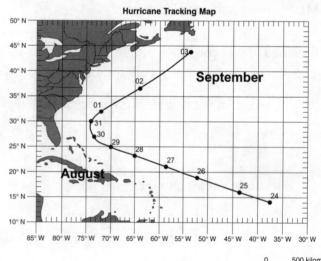

   c) Calculate the average daily rate of movement of the hurricane and label it with the proper units.

5. The highest elevation of Mt. Zembat in Alaska 40 years ago was measured at 7600 feet. Today the highest elevation is 7598 feet. What is the rate of change in elevation for this mountain.

   (1) 0.05 ft/yr
   (2) 0.6 ft/yr
   (3) 0.45 ft/yr
   (4) 20 ft/yr   5_____

6. A 25-gram sample of halite was placed in a jar with five other mineral samples and water. The jar was shaken vigorously for 5 minutes. The halite sample was then found to have a mass of 15 grams. What was the rate of weathering of the halite sample?

   (1) 0.50 g/min
   (2) 2.0 g/min
   (3) 3.0 g/min
   (4) 10.0 g/min   6_____

Base your answers to question 7 on the information and the accompanying map.

The eruption of Mt. St. Helens in 1980 resulted in the movement of volcanic ash across the northwestern United States. The movement of the ash at 1.5 km above sea level is shown as a shaded path on the map. The times marked on the path indicate the length of time the leading edge of the ash cloud took to travel from Mt. St. Helens to each location.

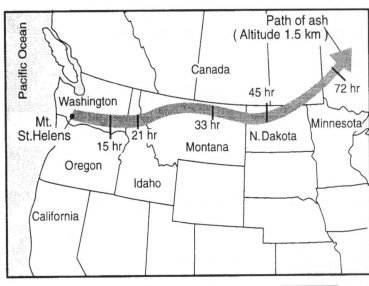

7. Calculate the average rate of movement of the volcanic ash for the first 15 hours, following the directions below.

   a) Write the equation used to determine the average rate of the volcanic ash movement.

   b) Substitute values into the equation.

   c) Solve the equation and label the answer with the correct units. _____

8. From 12 noon Thursday until 8 p.m. Thursday, the total amount of snowfall was 12 inches. Calculate the snowfall rate in inches per hour. _____

Rate of Change
Answers
Set 1

1. 2   The temperature of the water decreased from 60°C to 35°C, making the change in value 25°C. The time for the change was 10 minutes.

     Solution: Rate of change $= \dfrac{\text{change in value}}{\text{time}}$ $Rc = \dfrac{60\,°C - 35\,°C}{10\,\text{min}}$ $Rc = \dfrac{25\,°C}{10\,\text{min}} = 2.5\,°C/\text{min}$

2. 3   At 6 a.m., the temperature was 30°F. At noon the temperature rose close to 48°F. This makes the change in the temperature 18°F (48°F – 30°F). The time for this change was 6 hours.

     Solution: $Rc = \dfrac{48\,°F - 30\,°F}{6\,\text{hr}}$ $Rc = \dfrac{18\,°F}{6\,\text{hr}} = 3\,°F/\text{hr}$

3. 2   At minute 0, the temperature of the water in cup A was 90°C. Ten minutes later, the water cooled down to 77°C. This gives a change in value of 13°C.

     Solution: $Rc = \dfrac{90\,°C - 77\,°C}{10\,\text{min}}$ $Rc = \dfrac{13\,°C}{10\,\text{min}} = 1.3\,°C/\text{min}$

4 *a*)   Rate of movement (change) $= \dfrac{\text{change in value}}{\text{time}}$

   b) $Rc = \dfrac{2600 \text{ km}}{4 \text{ d}}$

   c) $Rc = 650$ km/d (± 50 km/d)

     Explanation: Use the edge of a piece of paper to mark off the distance the hurricane traveled during the four days. Use the given distance scale to determine distance in km.

Density

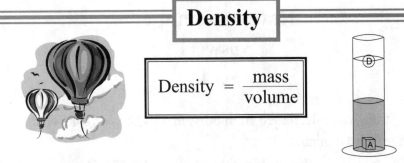

$$\text{Density} = \frac{\text{mass}}{\text{volume}}$$

Overview:

Density is an identifying property of matter. Calculating the density of an object, especially minerals, greatly assists in the identification of the object. Density is the mass per unit volume.

The Equation:

The density equation involves three variables: v, m, and d. When given any two variables, the third variable can be found. This can be shown by the accompanying "Density Triangle". From this triangle if one needs the equation for density, cover d and the answer is shown, $d = m/v$. The same procedure is used for mass, $m = (d)(v)$ and volume, $v = m/d$. The most common unit for density is g/cm³, which is the same as g/cc and g/mL.

If the density question involves a density graph, select any value along the mass axis and from this value move directly to the graph line. At this intersection point, read over to the volume axis to get its value.

At times the density value will change for a substance, especially for gases. When the volume of a gas changes due to temperature or pressure changes, its density is affected. For example, when the volume of a gas increases due to an increase in temperature or a decrease in pressure, the density will decrease. This is an inverse relationship: $\downarrow d = \frac{m}{v\uparrow}$ This is why a hot air balloon rises; the hot air expands the balloon ($v\uparrow$) causing its density to decrease. This action occurs constantly in the atmosphere due to the heating of the air by the Sun. This produces rising and sinking air currents, which may produce changes in the local weather. Remember, during these volume and density changes, the mass of the substance will remain the same.

Additional Information:

- Normally, solids are denser than liquids, and gases are the least dense. The exception to this is water. The solid phase, ice, is less dense than that of the liquid phase.

- When water changes to ice, the volume increases causing the density to decrease.

- Any object with a density value less than 1 g/cm³ will float when placed in water. An object with a density value higher than 1 g/cm³ will sink when placed in water.

- The same substance will have the same density value regardless of its size.

Diagram:

Density Measurements – A balance gives the mineral's mass, while the volume is obtained by using the displacement of water method. A submerged object displaces a volume of liquid equal to the volume of the object. The density of the mineral is 60 g/10 mL = 6 g/mL or 6 g/cm³ (1 mL = 1 cm³).

Mineral sample

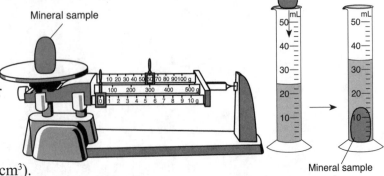

Mineral sample

Density

1. A rock sample has a mass of 16 grams and a volume of 8 cubic centimeters. When the rock is cut in half, what is the volume and density of each piece?

   (1) 8 cm³ and 0.5 g/cm³
   (2) 8 cm³ and 1.0 g/cm³
   (3) 4 cm³ and 2.0 g/cm³
   (4) 4 cm³ and 4.0 g/cm³   1_____

2. Liquid W was added to the graduated cylinder containing liquid C. Objects A and D were then dropped into the cylinder. Which statement is correct?

   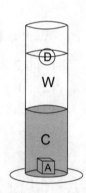

   (1) Liquid W is denser than liquid C and object D.
   (2) Liquid C is denser than liquid W and object A.
   (3) Liquid C is less dense than object A, but more dense than liquid W and object D.
   (4) Object A is denser than liquid C, but not as dense as liquid W and object D.   2_____

Note: Question 3 has only three choices.

3. As air on the surface of Earth warms, the density of the air

   (1) decreases
   (2) increases
   (3) remains the same   3_____

4. If the mass of a spinel crystal is 9.5 grams, what is the volume of this spinel crystal?

   Table 1

Gemstone Mineral	Composition	Hardness	Average Density (g/cm³)
emerald	$Be_3Al_2(Si_6O_{18})$	7.5–8	2.7
sapphire	Al_2O_3	9	4.0
spinel	$MgAl_2O_4$	8	3.8
zircon	$ZrSiO_4$	7.5	4.7

   (1) 0.4 cm³   (3) 5.7 cm³
   (2) 2.5 cm³   (4) 36.1 cm³   4_____

5. The graph below shows the relationship between mass and volume for three samples, A, B, and C, of a given material. What is the density of this material?

   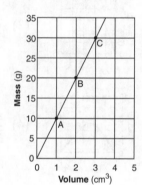

   (1) 1.0 g/cm³
   (2) 5.0 g/cm³
   (3) 10.0 g/cm³
   (4) 20.0 g/cm³

   5_____

6. The data table below, which shows the volume and mass of three different samples, A, B, and C, of the mineral pyrite.

Pyrite		
Sample	Volume (cm³)	Mass (g)
A	2.5	12.5
B	6.0	30.0
C	20.0	100.0

   State the mass of a 10.0-cm³ sample of pyrite.

   _____ g

7. The diagram below represents a solid object with a density of 3 grams per cubic centimeter. What is the mass of this object?

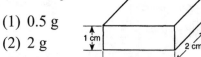

(1) 0.5 g
(2) 2 g
(3) 18 g
(4) 36 g

(Not drawn to scale)

7_____

8. Which graph best represents the relationship between mass and volume of a material that has a density of 5 grams per cubic centimeter?

(1)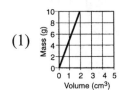

(3)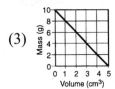

(2)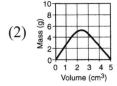

(4)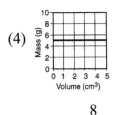

8_____

Note: Question 9 has only three choices.

9. When a substance is compressed the density would

(1) increase
(2) decrease
(3) remains the same

9_____

Base your answers to question 10 on the diagrams below. The diagrams represent two different solid, uniform materials cut into cubes *A* and *B*.

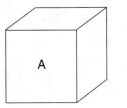

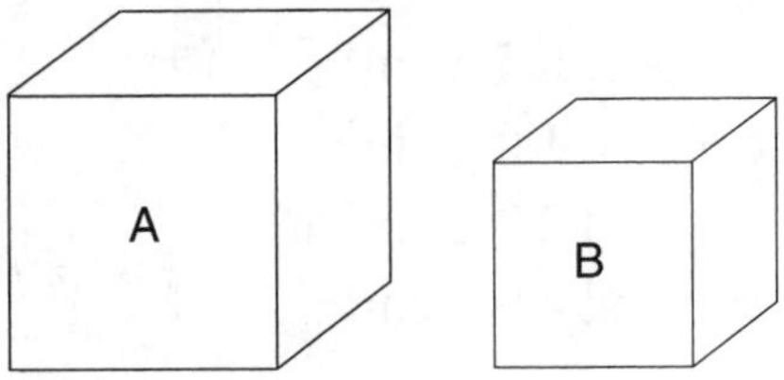

Mass of A = 320 g   Density of B = 3 g/cm³
Volume of A = 64 cm³   Volume of B = 27 cm³

(Not drawn to scale)

10. *a)* What is the density of cube *A*?

(1) 0.2 g/cm³
(2) 5.0 g/cm³
(3) 12.8 g/cm³
(4) 64.0 g/cm³

a_____

b) What is the mass of cube *B*?

(1) 3 g   (3) 27 g
(2) 9 g   (4) 81 g

b_____

Note: Question *c* has only three choices.

c) Assume cube *B* was broken into many irregularly shaped pieces. Compared to the density of the entire cube, the density of one of the pieces would be

(1) less
(2) greater
(3) the same

c_____

11. Explain how heat would change the density of a parcel of air.

_____

_____

1. 3   Density for the same object will be constant, regardless of the size. The mass of the rock sample is 16 g and the volume is 8 cm³. Substituting into the density equation and solving gives, $d = 16$ g$/8$ cm³ $= 2$ g/cm³. When the rock is cut in half, the volume is 4 cm³. Because the volume was cut in half, so will its mass, m = 8 g. Substituting these values and solving for density gives, $d = 8$ g$/4$ cm³ $= 2$ g/cm³. Notice, the density stays the same.

2. 3   Object A is the densest of all, because it sits on the bottom. Liquid C is denser than liquid W and object D, because both W and D float on top of liquid C.

3. 1   When a gas warms up, its volume increases, but the mass stays the same. This causes the density to decrease. This inverse relationship can be shown as: $v\uparrow, d\downarrow$. This density relationship can be observed when the air inside a hot air balloon gets heated causing the balloon to expand. This increase in volume causes the density to decrease, and the balloon starts to rise.

4. 2   From the density equation, $d = m/v$, the volume equation is $v = m/d$.

      Solution: $v = \dfrac{9.5\,\text{g}}{3.8\ \text{g}/\text{cm}^3} = 2.5$ cm³

5. 3   The graph line represents the density of samples A, B, and C of the same material. Because it is the same substance, the density of all samples will be the same. Choose any volume, and from this value go directly up until intersecting the graph line. From this intersection point, move to the left to get the mass. Substitute these values into the density equation to obtain the answer.

      For example: Using the volume of 3 cm³ and moving up to letter C,
                   the mass is 30 g,
      Solution: $d = 30$ g$/3$ cm³ $= 10$ g/cm³

6. Answer: 50 g

      Explanation: The density of pyrite will be the same regardless of its volume. First, the density needs to be obtained. Using the given information for Sample B:

                Equation: $d = m/v$
                Solution: $d = 30.0$ g$/6.0$ cm³ $= 5.0$ g/cm³

                Now solve for the mass of a 10 cm³ sample.
                Equation: $m = (d)(v)$
                Solution: $m = (5.0$ g/cm³$)(10.0$ cm³$) = 50$ g

Average Chemical Composition of Earth's Crust, Hydrosphere, and Troposphere

ELEMENT (symbol)	CRUST		HYDROSPHERE	TROPOSPHERE
	Percent by mass	Percent by volume	Percent by volume	Percent by volume
Oxygen (O)	46.10	94.04	33.0	21.0
Silicon (Si)	28.20	0.88		
Aluminum (Al)	8.23	0.48		
Iron (Fe)	5.63	0.49		
Calcium (Ca)	4.15	1.18		
Sodium (Na)	2.36	1.11		
Magnesium (Mg)	2.33	0.33		
Potassium (K)	2.09	1.42		
Nitrogen (N)				78.0
Hydrogen (H)			66.0	
Other	0.91	0.07	1.0	1.0

Overview:

Take a breath of air, and 99% of it is composed of nitrogen and oxygen. Dive into a lake, and you have slid by billions of water molecules consisting of two parts hydrogen and one part oxygen (H_2O). But take a shovel full of dirt, part of the Earth's crust, and you are holding many different compounds chemically made from the combination of less than 10 elements. Surprisingly, the crust's most abundant element is oxygen, and it readily combines with the Earth's crust second most abundant element, silicon, producing the compound silicone dioxide (SiO_2). This is the chemical formula for the mineral quartz, the second most abundant mineral in the Earth's crust. Elements, pure or chemically combined, produce minerals that make up the crust. Without these minerals, our planet would not be able to sustain life.

The Chart:

The crust is the only sphere that is divided into a mass and volume column. Oxygen has the greatest percentage in both mass and volume. Notice that the chart shows that eight elements make up over 99% of the total mass and volume of the crust. The hydrosphere, representing the water sphere, consists of 33% oxygen and 66% hydrogen, chemically combined to form the molecule H_2O. The troposphere is the first layer of our atmosphere, which we live in (see the Selected Properties of Earth's Atmosphere chart, page 172). This layer contains over 80% of all the gases of the total atmosphere, almost all of it being a mixture of 78% nitrogen and 21% oxygen. This is the air we breathe.

Additional Information:

- In the crust, oxygen and silicon chemically combine to form SiO_2, which is quartz the main component of sand. Silica, in the form of sand, is the main ingredient of glass.

- The other 1% of gases in the troposphere are mostly: water vapor, argon, and carbon dioxide.

- The most abundant mineral in the the crust is feldspar.

— Set 1 —

1. Earth's hydrosphere is best described as the

   (1) solid outer layer of Earth
   (2) liquid outer layer of Earth
   (3) magma layer located below
   Earth's stiffer mantle
   (4) gaseous layer extending several
   hundred kilometers from
   Earth into space   1 _____

2. What are the two most abundant elements
   by mass found in Earth's crust?

   (1) aluminum and iron
   (2) sodium and chlorine
   (3) calcium and carbon
   (4) oxygen and silicon   2 _____

3. Earth's troposphere, hydrosphere, and
   lithosphere contain relatively large
   amounts of which element?

   (1) iron   (3) hydrogen
   (2) oxygen   (4) potassium   3 _____

4. Which graph correctly represents the 3 most
   abundant elements, by mass, in Earth's crust?

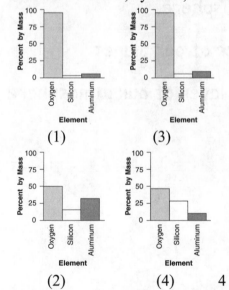

                                         4 _____

— Set 2 —

5. The most abundant element by
   volume in the hydrosphere is

   (1) oxygen   (3) nitrogen
   (2) hydrogen   (4) silicon   5 _____

6. Which metal is most abundant by
   mass in the crust?

   (1) iron   (3) magnesium
   (2) aluminum   (4) copper   6 _____

7. In which sphere would one find
   more total oxygen by volume?

   (1) the crust
   (2) the atmosphere
   (3) the hydrosphere
   (4) the mesosphere   7 _____

8. Which two elements would equal almost
   75% of the total mass of the crust?

   (1) O and Si   (3) Si and Al
   (2) O and Al   (4) Si and Fe   8 _____

9. The pie graph shows the elements comprising
   Earth's crust in percent by mass.

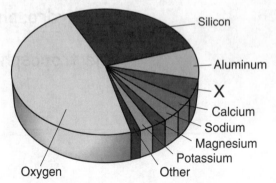

   Which element is represented by the letter X?
   (1) nitrogen   (3) iron
   (2) lead   (4) hydrogen   9 _____

Average Chemical Composition of Earth's Crust, Hydrosphere, and Troposphere

Answers

Set 1

1. 2 The hydrosphere includes all the water on and in our planet. The solid outer layer of our planet is the crust. The atmosphere, which includes the troposphere, is the gaseous layer that surrounds our planet.

2. 4 Open to the Average Chemical Composition of Earth's Crust, Hydrosphere, and Troposphere chart. In the the Percent by mass column, it shows that oxygen and silicon are the two most abundant elements in the crust.

3. 2 Oxygen is very abundant in all three spheres. The percentage of oxygen in each sphere is given in the chart. In the atmosphere, oxygen is found as a gas molecule, O_2. In the hydrosphere, oxygen is chemically combined with hydrogen, forming H_2O. In the crust (lithosphere), oxygen is chemically locked up in many compound-forming minerals.

4. 4 The chart shows the following composition by mass of the Earth's crust: oxygen 46.10%, silicon 28.20%, and aluminum 8.23%. These percentages match answer 4.

Remember:

- The crust is part of the lithosphere.
- The hydrosphere is the water of our planet.
- The troposphere is the first layer of our atmosphere.

Generalized Landscape Regions of New York State

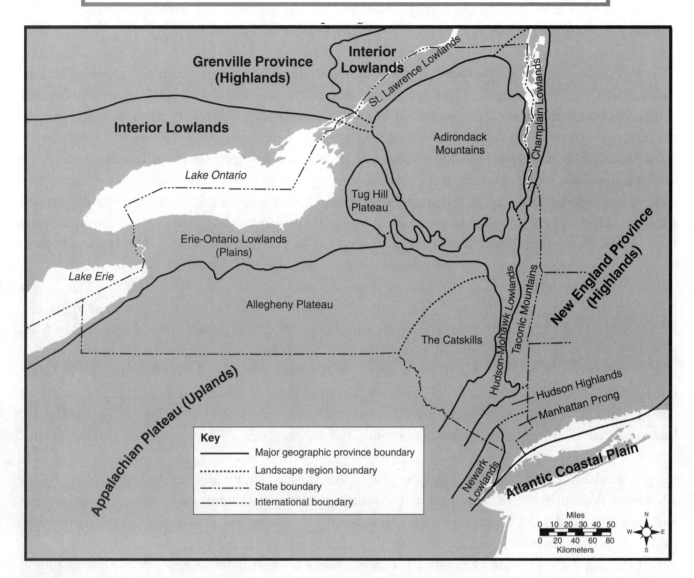

Overview:

Landscapes are classified as plains, plateaus, and mountains. New York State is one of the few states that has all three types. The development of a landscape is based primarily on the type of bedrock, climate factors, past and present geologic forces, and time. It takes tens of thousands or even millions of years for geologic forces to produce a landscape region. The Generalized Landscape Regions of NYS chart is often used with the Generalized Bedrock Geology of NYS and Geologic History of NYS charts to find related information. This might include arriving at the coordinates (latitude and longitude) of a landscape feature, or the type and age of bedrock and fossils found in specific landscape regions.

The relief of a landscape refers to the elevation differences found within the specific landscape. Mountainous landscapes normally will have the greatest relief, while plateaus usually exhibit medium relief and plains show very little change in relief. As one travels across NYS and is observant, it is easy to recognize a change in the landscape. This is especially true as one leaves the Atlantic Coastal Plain of Long Island and travels inland to various places within NYS.

The Map:

This map of NYS shows the location and extent of the different landscape regions throughout our state. It is important to know why certain regions are classified as a certain type of landscape.

New York State Mountain Landscapes – Tectonic plate movements are a dominant force producing mountains. The different types of mountains are folded, volcanic, fault-block, and dome. The bedrock structure in many mountain regions (but not all) will show folded or distorted layers at high elevation producing great relief. The Generalized Landscape Regions chart shows that NYS has two areas that are classified as mountain regions, the Adirondacks – a dome mountain region – and the Taconic Mountains – a folded mountain region. Both of these regions are very old and have undergone much erosion. This erosion has produced rounder peaks, not exhibiting the sharper peaks that are associated with younger mountains. The Adirondack Mountains consist of some of the oldest rocks found in North America, being Precambrian in age. The headwaters of many streams are located the Adirondacks. The Hudson River is the longest, flowing south and eventually discharging into the Atlantic Ocean.

New York State Plateau Landscapes – The bedrock structure of plateaus shows horizontal sedimentary rock strata that have been uplifted to produce the elevation needed to be considered a plateau region. Most people do not realize that NYS has any plateau regions. This is because NYS plateaus have been heavily eroded or dissected by glaciers and rivers, producing many large rolling hills with deep valleys. At times, this has caused people to mistakenly label certain plateau areas as mountains. The Landscape chart shows that the Catskills are part of the Allegheny Plateau. Even today, the Catskills are referred to as the Catskill Mountains, but their bedrock structure exhibits uplifted horizontal strata – a sure sign of a plateau region.

Notice that the Allegheny Plateau is an extension of the Appalachian Plateau that lies south of NYS. The Allegheny Plateau is NYS's largest plateau, consisting mostly of Devonian-age rocks. Locate the Tug Hill Plateau east of Lake Ontario. This plateau is noted for its large accumulation of snow. As the cold, moist winds travel eastward off Lake Ontario, they are forced upward by this plateau. This further cools the air causing the water vapor to condense, creating lake effect snowstorms when the air temperature is below 0°C.

New York State Plain Landscapes – Plains are relatively flat areas that show little relief. Plains that are adjacent to the ocean, are called coastal plains. Inland plains are referred to as interior lowlands. These flat or gentle rolling regions are home to many large cities. Here, people have settled next to major rivers or lakes. Extending beyond these cities is fertile farmland produced by the extensive weathering of the landscape bedrock. Long Island, located in the Atlantic Coastal Plain, geologically was developed by the deposition of great amounts of sediment transported by glaciers during the ice age. This geologic process makes Long Island the youngest landscape region in NYS.

Drainage Patterns – The local landscape will dictate the way streams flow producing certain stream patterns. For example, a large volcano will exhibit a pattern where streams radiate outward away from the center (see diagram). Flat landscapes usually have a main, large meandering stream with smaller streams (tributaries) flowing into it.

Page 44   **Generalized Landscape Regions**
                   of New York State

Additional Information:

- Landscapes located in a humid climate develop rounder hills, while landscapes located in an arid climate develop steeper, cliff-like features (e.g., Grand Canyon).

- Erosional forces such as glaciers, water, wind, etc., will slowly wear down landscapes. These forces are referred to as destructive forces.

- Plate collisions, volcanic eruption, and earthquakes will cause an increase in elevation that could change the landscape over time. These are referred to as constructive forces.

- A landscape that experiences a hot, moist climate will undergo much chemical weathering.

- A landscape that experiences a cold, moist climate will undergo much physical weathering.

Diagrams:

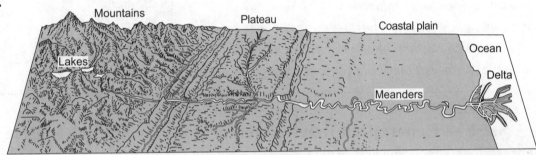

1. **Landscapes** – The diagram shows the general topography of the 3 different landscapes. Notice the rugged high relief and deep V-shaped valleys that mountainous landscape generally exhibit. The plateau landscape is being dissected by streams, eventually producing large rounded hills. The coastal plain, being relatively flat, usually will have a large meandering stream flowing to the ocean or to a large lake. Along both sides of the stream will be situated the floodplain – a fertile area, that as the name indicates, is prone to flooding. The delta is a fine grain depositional feature.

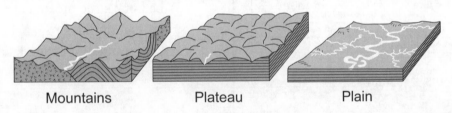

2. **Bedrock Structures** – These diagrams are the cross-sectional views of how the bedrock would appear for each landscape region. Mountainous regions will have folded/tilted layers with large areas of igneous and/or metamorphic rocks. Plateau regions will show elevated horizontal sedimentary bedrock layers. Coastal plains are very flat landscapes adjacent to the ocean, while lowlands are inland plains.

3. **A Glaciated Valley** – Originally, this mountainous landscape consisted of deep V-shaped valleys. The present U-shape valley was formed by the process of glaciation. Years ago, as the glaciers moved slowly down the valley, their massive size and weight scoured out the V-shaped valleys into U-shaped valleys as shown in the picture.

4. **Stream Drainage Pattern** – The topography of a landscape dictates how streams flow, producing a characteristic stream pattern. Notice how the high ridges separate the streams from joining. A steam that has successfully cut through a high ridge has produced a gap. The Delaware Water Gap is a famous one.

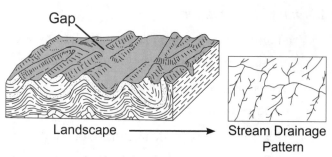

Landscape ⟶ Stream Drainage Pattern

5. **Cave Development** – Caves may affect the landscape topography of a region with the formations of sinkholes. Caves develop in limestone layers that are slowly dissolved by acidic water. The calcite minerals that make up limestone chemically react to acidic solutions. Over time, these hollowed-out "cavities" produce spectacular underground caverns that may connect for miles.

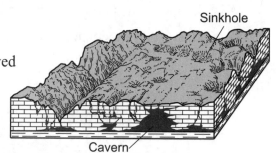

Sinkhole

Cavern

Set 1 — Generalized Landscape Regions of New York State

Note: For some questions, you will need to use the Generalized Bedrock Geology map on page 52.

1. The table below shows characteristics of three landscape regions, X, Y, and Z. Which terms, when substituted for X, Y, and Z, best complete the table?

Landscape Region	Relief	Bedrock
X	Great relief, high peaks, deep valleys	Many types, including igneous and metamorphic rocks, nonhorizontal structure
Y	Moderate to high relief	Flat layers of sedimentary rock or lava flows
Z	Very little relief, low elevations	Many types and structures

(1) X = mountains, Y = plains, Z = plateaus

(2) X = plateaus, Y = mountains, Z = plains

(3) X = plains, Y = plateaus, Z = mountains

(4) X = mountains, Y = plateaus, Z = plains   1_____

2. Which characteristics best distinguish one landscape region from another?

(1) human population density and types of environmental pollutants

(2) stream gradients and soil types

(3) bedrock structure and elevation of land surfaces

(4) composition of bedrock and variety of fossils   2_____

3. Which cross section best represents the general bedrock structure of New York State's Allegheny Plateau?

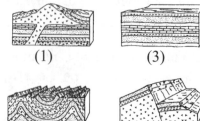

(1)   (3)

(2)   (4)   3_____

Generalized Landscape Regions of New York State

4. In which New York State landscape region is Niagara Falls located?

   (1) Tug Hill Plateau
   (2) St. Lawrence Lowlands
   (3) Allegheny Plateau
   (4) Erie-Ontario Lowlands   4_____

5. Which sequence shows the order in which landscape regions are crossed as an airplane flies in a straight course from Albany, New York, to Massena, New York?

   (1) plateau → plain → mountain
   (2) plateau → mountain → plain
   (3) plain → mountain → plain
   (4) mountain → plain → plateau

   5_____

6. Landscapes will undergo the most chemical weathering if the climate is

   (1) cool and dry (3) warm and dry
   (2) cool and wet (4) warm and wet

   6_____

7. Landscapes characterized by gentle slopes and meandering streams are most often found in regions with

   (1) steep mountain cliffs
   (2) sediment-covered bedrock
   (3) recently active faults and folds
   (4) high volcanic activity   7_____

8. Which New York State landscape region is composed mainly of metamorphosed surface bedrock?

   (1) Taconic Mountains
   (2) Allegheny Plateau
   (3) Atlantic Coastal Plain
   (4) Erie-Ontario Lowlands   8_____

9. Which New York State landscape region is located at 42° N 75° W?

   (1) Erie-Ontario Lowlands
   (2) the Catskills
   (3) Hudson-Mohawk Lowlands
   (4) Tug Hill Plateau   9_____

10. State the name of the New York State landscape region that includes location *A* shown in the diagram.

_____

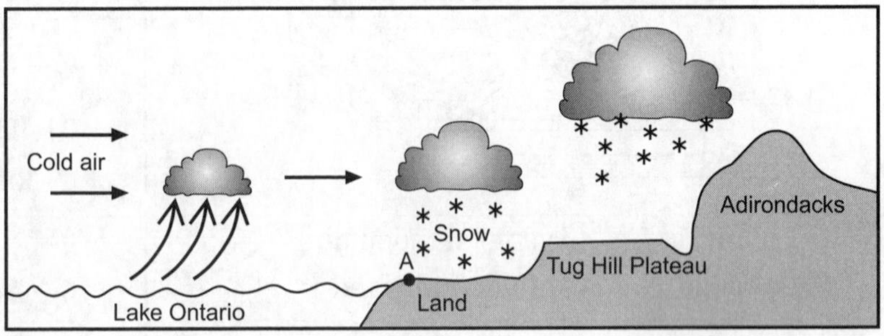

11. State why very heavy snowfall occurs in the Tug Hill Plateau region.

_____

12. Describe the most likely shape of a valley formed by erosion of a glacier.

_____

13. Which New York State landscape region is mostly composed of horizontal sedimentary bedrock at high elevations?

    (1) Hudson Highlands
    (2) Allegheny Plateau
    (3) Taconic Mountains
    (4) Atlantic Coastal Plain   13_____

14. The Erie-Ontario Lowlands of New York State are a part of which larger landscape region?

    (1) Interior Plains
    (2) St. Lawrence Lowlands
    (3) Allegheny Plateau
    (4) Appalachian Plateau   14_____

15. The diagram below shows a cross section of a portion of Earth's crust. Altitude is shown in meters above sea level. This landscape region is best classified as an eroded

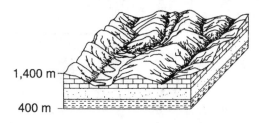

    (1) plain   (3) domed mountain
    (2) plateau   (4) folded lowland
                                         15_____

16. Which two locations are in the same New York State landscape region?

    (1) Albany and Old Forge
    (2) Binghamton and New York City
    (3) Massena and Mt. Marcy
    (4) Jamestown and Ithaca   16_____

17. The Catskills would best be described as

    (1) dissected plateau
    (2) part of the Hudson-Mohawk Lowlands
    (3) mountains with folded layers
    (4) volcanic in origin   17_____

18. On a yearly bases, agents of erosion are more active on

    (1) plains   (3) mountains
    (2) plateaus   (4) lowlands   18_____

19. Landscapes will undergo the most physical weathering if the climate is

    (1) cool and dry (3) warm and dry
    (2) cool and wet (4) warm and wet
                                         19_____

20. Tilted, slightly metamorphosed rock layers are typically found in which New York State landscape region?

    (1) Taconic Mountains
    (2) Atlantic Coastal Plain
    (3) Tug Hill Plateau
    (4) Erie-Ontario Lowlands   20_____

21. The photograph below shows an outcrop of horizontal rock layers in New York State.

Sandstone {
Shale {

Rock outcrops like this are most commonly found in which area of New York State?

    (1) Hudson Highlands
    (2) Adirondack Mountains
    (3) Atlantic Coastal Plain
    (4) Appalachian Plateau   21_____

22. The Catskills of New York State are best described as a plateau, while the Adirondacks are best described as mountains. Which factor is most responsible for the difference in landscape classification of these two regions?

(1) climate variations   (2) bedrock structure   (3) vegetation type   (4) bedrock age   22_____

23. The table below describes the characteristics of three landscape regions, A, B, and C, found in the United States.

Landscape	Bedrock	Elevation/Slopes	Streams
A	Faulted and folded gneiss and schist	High elevation Steep slopes	High velocity Rapids
B	Layers of sandstone and shale	Low elevation Gentle slopes	Low velocity Meanders
C	Thick horizontal layers of basalt	Medium elevation Steep to gentle slopes	High to low velocity Rapids and meanders

Which list best identifies landscapes A, B, and C ?

(1) A—mountain, B—plain, C—plateau   (3) A—plateau, B—mountain, C—plain
(2) A—plain, B—plateau, C—mountain   (4) A—plain, B—mountain, C—plateau   23_____

Base your answers to question 24 on the geologic cross section shown below. The cross section shows the surface of a landscape region in the southwestern United States and indicates the age, type, and thickness of the bedrock.

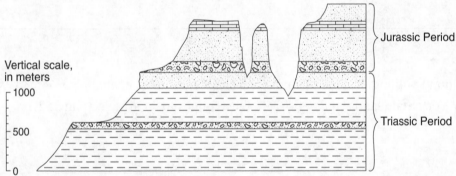

24. State *two* characteristics shown in the cross section that supports the idea that this region is correctly classified as a plateau landscape.

1)_____

2)_____

25. Give the landscape regions that the Genesee river flows over.

_____

26. Identify the landscape region in which the Finger Lakes are located. _____

Base your answer to question 27 on the map below. The map shows the approximate area in a portion of North America where some sedimentary rock layers composed of gypsum, halite, and potassium salt minerals are found in Earth's crust.

Mineral Deposits

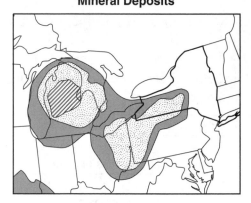

Key

▨ Gypsum	⬚ Gypsum and halite
▨ Gypsum, halite, and potassium salts	

27. Identify one New York State landscape region in which deposits of gypsum and halite are commonly found.

_____

28. New York State's Adirondacks are classified as a mountain landscape region. Describe one bedrock characteristic and one land surface characteristic that were used to classify the Adirondacks as a mountain landscape region.

Bedrock characteristic – _____

Land Surface characteristic – _____

29. Part of which generalized New York State landscape region is drained by the Susquehanna River and its tributaries?

_____

30. Chrysotile is found with other minerals in New York State mines located near 44° 30′ N, 74° W. In which New York State landscape region are these mines located? (See page 52.)

_____

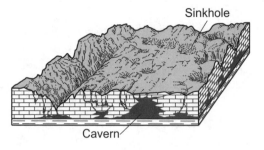

31. *a)* Why did these caverns and sinkholes form?_____

_____

b) What type of weathering was primarily responsible for the development of theses caves?

_____

Generalized Landscape Regions of New York State

1. 4   Landscape region *X*, having high peaks and deep valleys, describes a mountainous region. The bedrock of most mountains consists of igneous and metamorphic rocks that are resistant to erosion. Plateaus (*Y*) contain horizontal layers of sedimentary rocks having medium relief. Plains (*Z*) are generally flat, eroded landscapes at low elevation.

2. 3   The bedrock structure of an area is the primary reason for the type of landscape that exists in a region. Elevation of the land surface is another identifying property used in the classification of a landscape region.

3. 3   The underlying bedrock of plateaus shows horizontal sedimentary layers, as shown by diagram 3. The other diagrams show volcanism, folding, and faulting.

4. 4   On the Generalized Bedrock Geology map, locate Niagara Falls – in the upper western part of NYS. From the Generalized Landscape Regions of NYS map, Niagara Falls is in the Erie-Ontario Lowlands.

5. 3   On the Generalized Bedrock Geology map, locate Albany (eastern section) and Massena (northern section). Locate these positions on the Generalized Landscape Regions map. Albany is part of the Hudson-Mohawk Lowlands. Lowlands are classified as plains. Flying north to Massena, one would past over the Adirondack Mountains and then reach the plains of the St. Lawrence Lowlands.

6. 4   Chemical weathering is favored by a warm or hot climate with much rainfall.

7. 2   Gentle slopes and meandering streams are found on plains. Stream velocity decreases on plains, causing streams to deposit their sediments on the existing bedrock.

8. 1   Using the Landscape map, locate the Taconic Mountains. Now find this area on the Bedrock Geology map. Matching the shaded pattern of the Taconic Mountain bedrock to the given key shows that the bedrock in this area is mostly Taconic Sequence – "Slightly to intensely metamorphosed rocks."

9. 2   Open to the Generalized Bedrock Geology map. The latitude and longitude readings are located on the edges of the state map. The intersection of 42° N and 75° W is just west of Slide Mt. Using the Landscape Regions of NYS map, this position is located in the Catskills.

10. Answer: Erie-Ontario Lowlands (plains)

    Explanation: Open to the Generalized Landscape Regions of NYS chart. The eastern shore of Lake Ontario is adjacent to the Erie-Ontario Lowlands. This is where location *A* is.

11. Answer: As the moist air is forced upward by the topography, its temperature cools and reaches the dewpoint temperature, producing condensation and forming clouds. Eventually, heavy lake-effect snowstorms may develop.

12. Answer: U-shaped valley

    Explanation: See diagram 3 page 45.

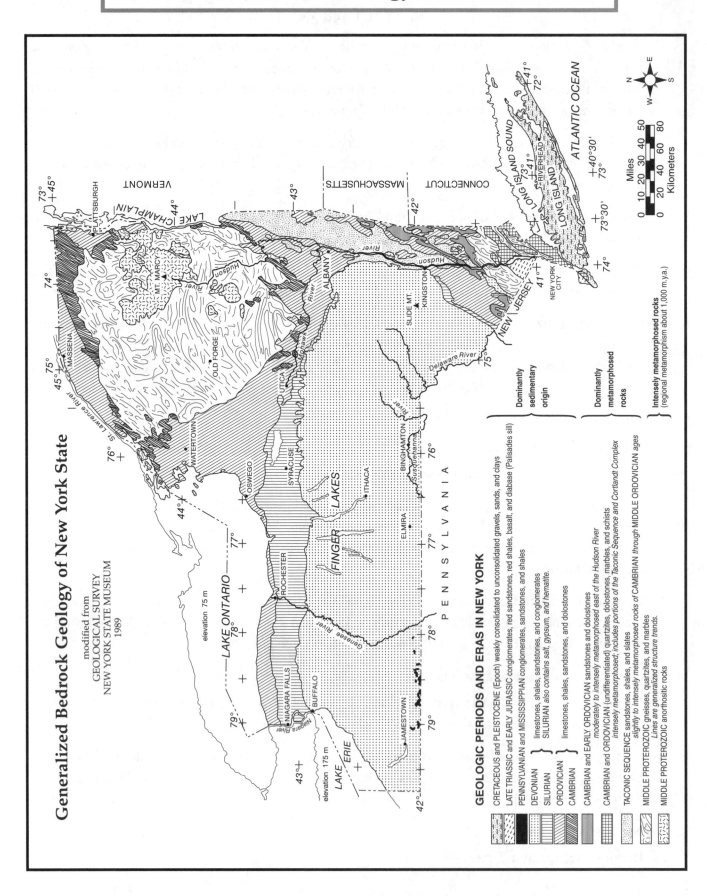

Generalized Bedrock Geology of New York State

modified from
GEOLOGICAL SURVEY
NEW YORK STATE MUSEUM
1989

GEOLOGIC PERIODS AND ERAS IN NEW YORK

CRETACEOUS and PLEISTOCENE (Epoch) weakly consolidated to unconsolidated gravels, sands, and clays

LATE TRIASSIC and EARLY JURASSIC conglomerates, red sandstones, red shales, basalt, and diabase (Palisades sill)

PENNSYLVANIAN and MISSISSIPPIAN conglomerates, sandstones, and shales

DEVONIAN limestones, shales, sandstones, and conglomerates

SILURIAN *also contains salt, gypsum, and hematite.*

ORDOVICIAN limestones, shales, sandstones, and dolostones

CAMBRIAN

CAMBRIAN and EARLY ORDOVICIAN sandstones and dolostones
moderately to intensely metamorphosed east of the Hudson River

CAMBRIAN and ORDOVICIAN (undifferentiated) quartzites, dolostones, marbles, and schists
intensely metamorphosed; includes portions of the Taconic Sequence and Cortlandt Complex

TACONIC SEQUENCE sandstones, shales, and slates
slightly to intensely metamorphosed rocks of CAMBRIAN through MIDDLE ORDOVICIAN ages

MIDDLE PROTEROZOIC gneisses, quartzites, and marbles
Lines are generalized structure trends.

MIDDLE PROTEROZOIC anorthositic rocks

Dominantly sedimentary origin

Dominantly metamorphosed rocks

Intensely metamorphosed rocks
(regional metamorphism about 1,000 m.y.a.)

Overview:

This map shows the type and the geologic ages of the surface bedrock that is found in NYS. Most of the time one does not see the local bedrock, because it is usually covered by vegetation, soil, concrete, or blacktop. But if one could scrape off these layers, one would eventually reach this solid rock layer. From exposed outcrops, geologists have mapped the location and extent of the different types of bedrock found throughout NYS. The bedrock types have been classified according to their geologic age.

The Map:

Key Area – The Geologic Periods and Eras in New York – The key lists 12 different geologic divisions, starting with the youngest – the Pleistocene Epoch – and ending with the oldest – the Middle Proterozoic – with each one representing a bedrock layer. As one goes down this key, the age of the bedrock increases. To the right of these geologic divisions are the type of rocks that make up the bedrock. For example, the Devonian bedrock layers are dominantly sedimentary rocks consisting of limestones, shales, sandstones, and conglomerates. Notice that the last two divisions, both Middle Proterozoic, are part of the very old Precambrian Eon. The key also shows that the Middle Proterozoic rocks are intensely metamorphosed. These rocks being metamorphic and so old, fossils would not be found in them. The Adirondack Mountains (see Landscape Regions chart, page 43) consist mostly of the Middle Proterozoic bedrock.

Bedrock Geology of NYS Map – The map gives the location of the different types of surface bedrock based on their geologic ages. Once the geologic name of the bedrock is known, and using the Geologic History of NYS chart (page 106), one can get the age range of the bedrock and can identify specific fossils that lived during that time period. For example, the map shows that the bedrock around the Finger Lakes is Devonian in age. The Geologic History of NYS chart states that this Devonian bedrock is 359 to 416 million years old. At the bottom of this chart are diagrammed Devonian index fossils that can be expected to be found in this age bedrock and therefore around the Finger Lakes region. Using the Bedrock Geology chart, north of the Finger Lakes region, the Silurian layer is found. The key shows that the Silurian layer is older than the Devonian layer. Thus, where one finds Devonian bedrock, usually Silurian bedrock would be located under it. Using this method, the Ordovician layer would be under the Silurian layer, and so forth.

Latitude and Longitude Coordinates – One can locate latitude numbers using the right and left sides of the map. The 42° represents the latitude reading of 42° N. The longitude numbers can be found along the bottom and top of the map. The 76° represents the longitude reading of 76° W. Halfway between the latitude and longitude readings is a line marking the 30′ (minutes) position. Remember, that one degree of latitude or longitude is subdivided into 60 minutes. Example: Find the coordinates of Ithaca. First find Ithaca's latitude. Using a straight edge, the map shows that Ithaca is higher that 42° N but slightly less than 42° 30′ N, being very close to 42° 20′ N. Its longitude is very close to 76° 30′ W, giving the coordinates of Ithaca as 42° 20′ N, 76° 30′ W. (See Latitude, Longitude, and Time Zones chapter, page 216.)

The Angle of Polaris – The altitude of *Polaris* (the North Star) is equal to the observer's northern latitude. If you live in Niagara Falls, with the latitude very close to 43° N, the altitude of *Polaris* would be 43°. As one travels north in the Northern Hemisphere, the angle of *Polaris* increases.

Other Features – New York State's major cities are shown along with major lakes and rivers. The longest river in NYS is the Hudson River. Its headwaters are located in the Adirondack Mountains, and it flows south passing over many different bedrock layers on its way to the Atlantic Ocean. Locate the Genesee River. It flows north over three Paleozoic layers until it discharges into Lake Ontario. Notice that the elevations of both Great Lakes are given on this map. One of the smallest rivers shown is the Niagara River. This river's water comes from the overflow of Lake Erie. As it flows to Lake Ontario, it drops over a large escarpment (cliff), producing the famous Niagara Falls. Mount Marcy, located in the Adirondacks, has the highest elevation in our state at 5,343 feet. Slide Mountain, located in the Catskills, is the highest peak in this region. On the bottom right of this map is the distance scale in miles and kilometers.

Watersheds – A watershed is a region or area that is drained by a major river and its smaller tributaries. When surface runoff occurs, small streams collect this water and move it into larger streams that are at a lower elevation. These stream usually flow into a major river, which will eventually discharge into a large lake, sea, bay, or ocean. In NYS, there are many different watersheds. These are separated from each other by the topography of the land. The higher elevation region that separates one watershed from another is called a *divide*. These watersheds provide valuable drinking water for many cities and support many habitats. For these reasons, and many others, watersheds must be protected from pollution and mismanagement. The map below shows the boundaries of the different watersheds found in NYS. Studying it, one can see a general pattern of the locations of the watersheds when viewed with the Generalized Bedrock Geology and the Landscape Regions maps.

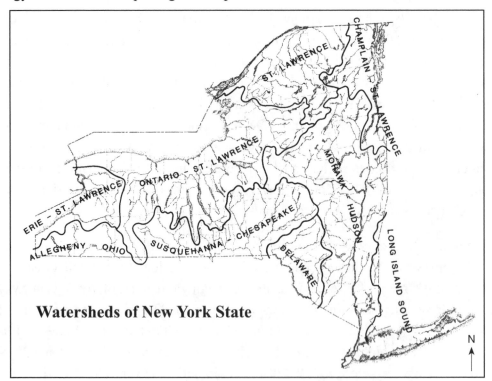

Watersheds of New York State

Additional Information:

NYS has a small section of Mesozoic rocks (dinosaur layers) labeled in the key area as Late Triassic and Early Jurassic Period. This small area can be found north of NYC, bordering New Jersey. The footprint of the raptor Coelophysis was found here. Locate its skeleton and you might become very rich. Happy hunting.

**Generalized Bedrock Geology
of New York State**

Diagrams:

1. **Bedrock Layers** – The surface bedrock is of Devonian age. As erosion continues, the Silurian layer will become exposed as surface bedrock. A drill core taken from location *A* reveals that the Ordovician layer is missing. This layer must have been eroded away sometime in the past producing an unconformity.

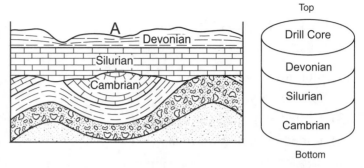

2.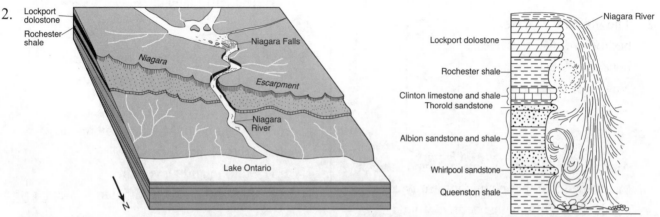

Resistant Bedrock – The Niagara escarpment is a cliff-like structure. Its surface bedrock consists of Lockport dolostone that is more resistant to weathering than the layers below. These differences in weathering rates has produced Niagara Falls. The cross-section diagram shows the Niagara River flowing over the tougher dolostone layer producing Niagara Falls. Eventually the Lockport dolostone will be undercut by the weaker, less resistant layers and break off in large chucks tumbling to the base of the Niagara River.

3. **Watersheds** – Surface runoff flows downward by the force of gravity, but it is the local landscape and bedrock structures that direct the flow of the surface runoff. The surrounding topography will cause water to be confined to a specific area that drains the surface water. This is it's watershed. Divides (mountain peaks and ridgelines) separate one watershed from another.

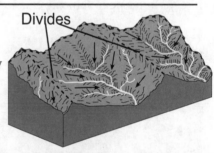

4.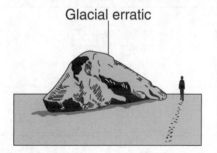

Glacial Action on Bedrock – Large boulders (erratics) whose composition does not match the local bedrock and bedrock that was grooved and polished are clues of glacial action. Glaciation on NYS bedrock proved that the ice sheet moved southward during the ice ages.

1. Bedrock in the area of Binghamton, New York, consists of

   (1) plutonic igneous rock
   (2) sedimentary rock layers
   (3) faulted and tilted volcanic rock
   (4) folded metamorphic rock   1 _____

2. The diagram represents bedrock of different ages beneath a location in New York State.

   Assuming that the rock layers have not been overturned and that no unconformity exists, at which location is this bedrock found?

   (1) Albany   (3) Elmira
   (2) Old Forge   (4) Oswego   2 _____

3. Which types of surface bedrock are most likely found near Jamestown, New York?

   (1) slate and marble
   (2) quartzite and granite
   (3) shale and sandstone
   (4) schist and gneiss   3 _____

4. Gneiss and quartzite rocks are found in the metamorphic surface bedrock in which New York State landscape region?

   (1) Catskills
   (2) Adirondacks
   (3) Erie-Ontario Lowlands
   (4) Tug Hill Plateau   4 _____

5. What is the age of most of the surface bedrock found in New York State at a latitude of 45° N?

   (1) Precambrian Middle Proterozoic
   (2) Triassic and Jurassic
   (3) Silurian and Devonian
   (4) Cambrian and Ordovician   5 _____

6. At which New York State location will an observer most likely measure the altitude of *Polaris* as approximately 42°?

   (1) Jamestown (3) New York City
   (2) Plattsburgh (4) Oswego   6 _____

7. During which geologic epoch does the New York State rock record consist of weakly consolidated to unconsolidated sediments?
   (1) Early Permian (3) Late Cretaceous
   (2) Early Jurassic (4) Pliocene   7 _____

8. State the latitude and longitude coordinates of Mt. Marcy, New York. The units and compass directions must be included in your answer. _____

9. Give the name of the surface bedrock found at Slide Mt.

   and identify the geologic age (in millions of years). _____ _____mya

10. Give the geologic age sequence of the surface bedrock from Ithaca to Watertown.

_____

11. Which New York State river flows generally southward?

    (1) St. Lawrence River
    (2) Mohawk River
    (3) Genesee River
    (4) Hudson River   11_____

12. The approximate latitude of Utica, New York, is

    (1) 43°05′ N   (3) 75°15′ E
    (2) 43°05′ S   (4) 75°15′ W   12_____

13. Near which location in New York State would a geologist have the greatest chance of finding dinosaur footprints in the surface bedrock?

    (1) 41° 10′ N latitude, 74° W longitude
    (2) 42° 10′ N latitude, 74° 30′ W longitude
    (3) 43° 30′ N latitude, 76° W longitude
    (4) 44° 30′ N latitude, 75° 30′ W longitude

    13_____

14. Which New York State landscape region is composed mostly of intensely metamorphosed surface bedrock?

    (1) Hudson Highlands
    (2) Allegheny Plateau
    (3) Atlantic Coastal Plain
    (4) Erie-Ontario Lowlands   14_____

15. Which two cities would have the same surface fossils?

    (1) Syracuse and Watertown
    (2) Old Forge and Massena
    (3) Elmira and Binghamton
    (4) Rochester and Elmira   15_____

16. At which latitude and longitude in New York State would a salt mine in Silurian-age bedrock most likely be located?

    (1) 41° N 72° W   (3) 44° N 74° W
    (2) 43° N 77° W   (4) 44° N 76° W

    16_____

17. What is the age of the most abundant surface bedrock in the Finger Lakes region of New York State?

    (1) Cambrian   (3) Pennsylvanian
    (2) Devonian   (4) Permian   17_____

18. The Generalized Bedrock Geology Map of New York State provides evidence that water flows from Lake Erie into Lake Ontario by showing that Lake Ontario

    (1) is north of Lake Erie
    (2) has lower surface elevation
        than Lake Erie
    (3) has a larger surface area
        than Lake Erie
    (4) is deeper than Lake Erie   18_____

19. Which river is a tributary branch of the Hudson River?

    (1) Delaware River   (3) Mohawk River
    (2) Susquehanna River   (4) Genesee River

    19_____

20. In which New York State landscape region is most of the surface bedrock composed of metamorphic rock?

    (1) Adirondacks (3) Erie-Ontario Lowlands
    (2) Catskills   (4) Tug Hill Plateau

    20_____

21. New York State bedrock of which age contains salt, gypsum, and hematite?

(1) Cambrian (2) Devonian (3) Mississippian (4) Silurian 21_____

22. Bedrock of which four consecutive geologic periods is best preserved in New York State?

(1) Cambrian, Ordovician, Silurian, Devonian
(2) Devonian, Carboniferous, Permian, Triassic
(3) Permian, Triassic, Jurassic, Cretaceous
(4) Jurassic, Cretaceous, Tertiary, Quaternary 22_____

23. The accompany diagram shows an index fossil found in surface bedrock in some parts of NYS. In which NYS landscape region is this gastropod fossil most likely found in the surface bedrock?

(see page 107)

Maclurites

(1) Tug Hill Plateau (3) Adirondack Mountains
(2) Allegheny Plateau (4) Long Island 23_____

24. The accompanying map shows major streams in the New York State area. The bold lines mark off sections A through I within New York State. The best title for the map would be

(1) "Tectonic Plate Boundaries in New York State"
(2) "Bedrock Geology Locations of New York State"
(3) "Landscape Regions of New York State"
(4) "Watershed Areas of New York State"

24_____

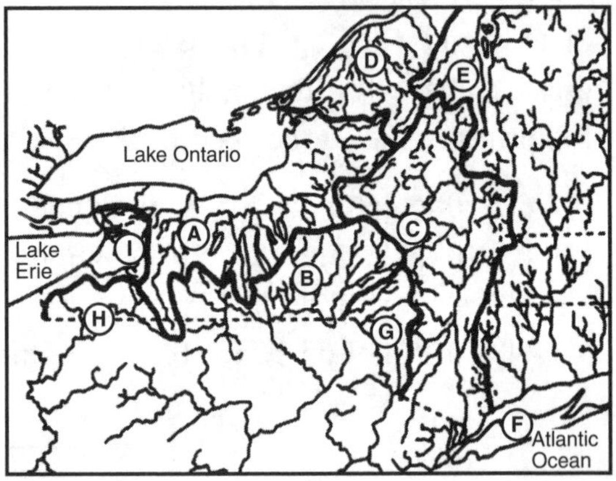

25. As one flies from Elmira to Old Forge, what is true about the age of the bedrocks one passes over?

_____

26. Explain why one would have a better chance of finding the fossil Phacops, a trilobite, in the surrounding bedrock of Ithaca, than in the surrounding bedrock of Watertown. (See page 106.)

_____

27. Give a statement comparing the height of Polaris at Watertown to the height of Polaris at Binghamton.

_____

28. Which NYS geologic feature has surface bedrock that consists mainly of anorthositic rock?

_____

**Generalized Bedrock Geology
of New York State**

1. 2   Open to the Bedrock Geology of NYS chart and locate Binghamton in the south central part of NYS. The Binghamton area is situated on Devonian bedrock. To the right of the Devonian key it states: "Dominantly Sedimentary Origin."

2. 3   The correct city must be located on Devonian bedrock, because this is the top layer. Locating each city, only Elmira is situated on Devonian bedrock. If one were to drill downward through the Devonian bedrock, the next older layer would be the Silurian bedrock, then the Ordovician bedrock, and then the Cambrian layer.

3. 3   Jamestown's bedrock is of Devonian age. The key gives the composition of Devonian bedrock as limestone, shale, sandstone and conglomerates.

4. 2   Gneiss and quartzite are metamorphic rocks. Open to the Generalized Landscape of NYS chart and notice the location of the Adirondack Mountains. Locate this area on the Bedrock chart. The bedrock key area shows that the Adirondack's bedrock is mostly Middle Proterozoic in age. To the far right of this bedrock layer it states: "Intensely Metamorphosed Rocks."

5. 4   The latitude numbers are located on the right side of the map. Locate the 45° N latitude line. Using the Geological Periods and Eras key, the bedrock on this latitude line is Cambrian and Ordovician.

6. 1   The angle of Polaris, measured up from the northern horizon, is equal to the observers latitude. This can be restated as: An observer's latitude is equal to the angle of Polaris. Locate Jamestown in the west central section of NYS. This town is located just north of the 42° N latitude line. Thus looking north, Polaris would be very close to 42° high in the night sky.

7. 3   In the key area, the first period is Cretaceous in age. Here it states, "weakly consolidated to unconsolidated gravels, sands, and clays."

8. Answer: 44° 10′ N, 74° W

   Explanation: Using a straight edge, line up the coordinates found on the sides of the map. Mt. Marcy is higher than 44° N, and lower than 44° 30′ N latitude marks found on the right and left side of the map. An acceptable answer for latitude is one that is close to 44° 10′ N. The 74° W longitude marking on the top and bottom of the map runs directly through Mt. Marcy.

9. Answer: Geologic name is Devonian.
   Geologic age is 359 to 416 million years ago.

   Explanation: Slide Mt. is located in the Catskills, and its surface bedrock is Devonian. On the Geologic History of NYS chart (see page 106), find the time line next to the Epoch column and move down to the Devonian Period. The geologic age span is given at the start and at the end of this time period.

10. Answer: Devonian – Silurian – Ordovician

    Explanation: Match the different surface bedrock layers to their geologic periods in the key.

Surface Ocean Currents

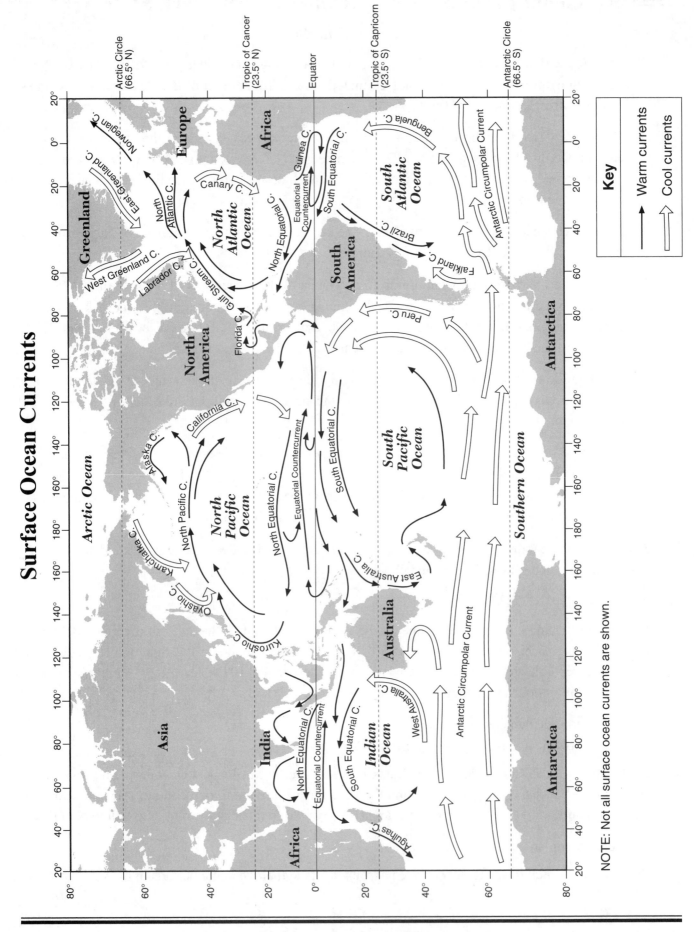

Key
→ Warm currents
⇒ Cool currents

NOTE: Not all surface ocean currents are shown.

Surface Ocean Currents

Overview:

Surface ocean currents are shallow currents that are driven by wind. Our atmosphere contains large prevailing wind belts located in specific zones (see the Planetary Wind and Moisture Belts in the Troposphere chart, page 177). These wind belts tend to blow in one direction, forcing the surface ocean water to move in the same direction, thus setting up these currents. Benjamin Franklin, studying logbooks of ocean sailing ships, realized that sailing with these currents quickened the journey. He encouraged sailors leaving Europe, heading to America, to first sail south then west to take advantage of the surface currents that move in those directions. Ships departing from America heading to Europe should sail with the Gulf Stream Current that flows northeast toward Europe. The paths of these ocean currents are affected by landmasses, the rotation of the Earth, and when they enter a different wind belt.

"Here comes another big one, Roy, and here—we—gooooooowheeeeeeeoooo!"

The Map:

The key shows the classification of shallow ocean currents: warm currents (shown by a dark arrow) and cool currents (shown by an outline arrow). As expected, currents that originate near the equator are warm currents, and those originating in the high latitudes are cool currents. Locate the California and the Alaska currents. Contrary to what one would think, the California Current is a cool current, and the Alaska Current is a relatively warm current. The Gulf Stream Current flows northeast and eventually becomes the North Atlantic Current. This warm current is a major factor affecting the climate of Iceland and the coastal areas of northwestern Europe. Probably the most feared water to sail in is the area of the Antarctic Circumpolar Current. This area is noted for fast winds that produce this strong ocean current with notoriously large waves. Notice that, in this region, there is no interruption by any landmass, causing both the wind and the resulting ocean current to maintain their strength and speed as they circle the globe.

Additional Information:

- The map shows that ocean surface currents tend to curve. The Coriolis effect causes this curvature path and is produced by the rotation of the Earth. Objects that travel long distances over the Earth's surface, like ocean currents and wind, tend to curve or get deflected to the right in the Northern Hemisphere and to the left in the Southern Hemisphere. The Coriolis effect is recognized as a proof of the Earth's rotation.

- El Niño is a natural event which temporarily slows or reverses surface ocean currents in the equatorial region of the Pacific Ocean. This occurrence, causes wide-spread disruption of normal weather patterns for many regions of the world (see page 179, diagram 3).

— Set 1 —

1. Which ocean current flows northeast along the eastern coast of North America?

    (1) Gulf Stream
    (2) North Equatorial
    (3) California
    (4) Labrador   1_____

2. The California Ocean Current, which flows along the west coast of North America, is a

    (1) cool current, flowing north
    (2) cool current, flowing south
    (3) warm current, flowing north
    (4) warm current, flowing south   2_____

3. Surface ocean currents curve to the right in the Northern Hemisphere because

    (1) the Moon's gravity curves ocean currents
    (2) the Moon travels in an orbit around Earth
    (3) Earth spins on its axis
    (4) Earth travels in an orbit around the Sun   3_____

4. Surface ocean currents located at 40° south latitude, 90° west longitude generally flow toward the

    (1) northeast   (3) southwest
    (2) southeast   (4) west   4_____

5. Which surface ocean current is located at 30° N, 75° W?

    _____

6. What direction does the East Greenland Current flow?

    _____

— Set 2 —

7. Which current is a cool ocean current that flows completely around Earth?

    (1) Antarctica Circumpolar Current
    (2) Gulf Stream Current
    (3) North Equatorial Current
    (4) California Current   7_____

8. Surface ocean currents resulting from the prevailing winds over the oceans illustrate a transfer of energy from

    (1) lithosphere to atmosphere
    (2) hydrosphere to lithosphere
    (3) atmosphere to hydrosphere
    (4) stratosphere to troposphere   8_____

9. Which two currents are considered to be warm currents?

    (1) Brazil Current and
        Antarctic Circumpolar Current
    (2) Kuroshio Current and the
        California Current
    (3) Alaska Current and the Brazil Current
    (4) North Atlantic Current and
        the Benguela Current   9_____

10. Which current has a major factor on the climate of Western Europe?

    (1) Labrador Current
    (2) Canary Current
    (3) Gulf Stream Current
    (4) North Atlantic Current   10_____

11. Identify the ocean current off the northeast coast of Australia that most affects the climate of Ambae Island.

    _____

12. The Canary Current along the west coast of Africa and the Peru Current along the west coast of South America are both

(1) warm currents that flow away from the Equator
(2) warm currents that flow toward the Equator
(3) cool currents that flow away from the Equator
(4) cool currents that flow toward the Equator   12_____

13. Which surface ocean current transports warm water to higher latitudes?

(1) Labrador Current
(2) Falkland Current
(3) Gulf Stream
(4) West Wind Drift   13_____

14. Identify one warm and one cool ocean current that affect the climate of Iceland.

_____ _____

15. Which diagram correctly represents the curving of Earth's ocean currents and prevailing winds due to the Coriolis effect?

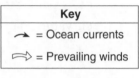

Key
= Ocean currents
= Prevailing winds

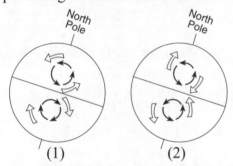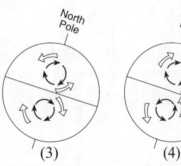

(1)   (2)   (3)   (4)   15_____

16. The map below shows four coastal locations labeled *A*, *B*, *C*, and *D*.

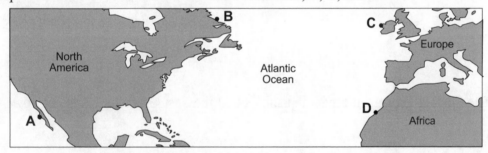

The climate of which location is warmed by a nearby major ocean current?
(1) *A*   (2) *B*   (3) *C*   (4) *D*   16_____

17. Identify by name the surface ocean current that cools the climate of locations on the western coastline of North America. _____

18. What controls the direction of movement of most surface ocean currents?

_____

19. Identify the cool surface ocean current that prevents the formation of coral reefs in the shallow waters along the western coast of South America. _____

Surface Ocean Currents

Answers

Set 1

1. 1 The Surface Ocean Currents map shows that along the eastern coast of North America, the warm Gulf Stream Current flows northeast. Eventually it becomes the North Atlantic Current.

2. 2 The California Current is shown by an outline arrow flowing south. From the key, this is a cool current.

3. 3 An object that travels long distances on a rotating planet will experience a deflection or curvature of its path. Surface ocean currents and planetary winds are two examples that undergo this curvature. This is a proof of the Earth's rotation and is called the Coriolis effect.

4. 1 The position of 40° S, 90° W is off the western coast of Peru. At this location, the outline arrow of the cool Peru Current is heading northeast along the west coast of South America.

5. Answer: Gulf Stream Current

   Explanation: Locate the 75° W longitude value at the bottom of the map. Follow this longitude line upwards, stopping where the estimated 30° N latitude line crosses it. At this position, the Gulf Stream flows.

6. Answer: southwest

   Explanation: The Surface Ocean Currents map shows that the East Greenland Current flows southwest.

Remember:

Surface ocean currents are moved by the prevailing winds of that region.

Tectonic Plates

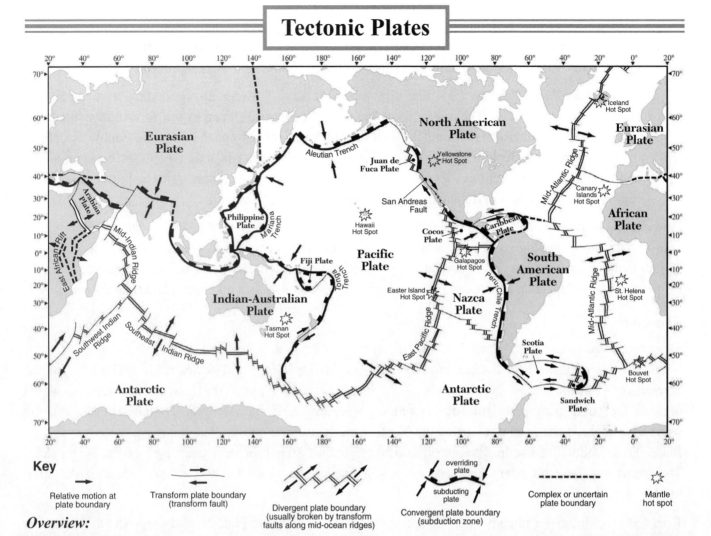

Key

→	→ ←	⊬⊬⊬	overriding plate / subducting plate	----------	✦
Relative motion at plate boundary	Transform plate boundary (transform fault)	Divergent plate boundary (usually broken by transform faults along mid-ocean ridges)	Convergent plate boundary (subduction zone)	Complex or uncertain plate boundary	Mantle hot spot

Overview:

The Earth's lithosphere consists of major and minor tectonic plates that are being pushed/pulled by forces within the Earth. The net results of this "bumper car" action are earthquakes, volcanoes, and mountain ranges, along with many other geologic processes that have their origin with the movement and collisions involving these plates.

Tectonic Plates and Boundaries:

Tectonic plates consist of two types: (1) oceanic plate and (2) continental plate. The oceanic plate composition is basaltic, while the continental plate is granitic. The oceanic plate is thinner but denser than the continental plate. Both plates are part of the lithosphere, the solid outer part of the Earth, consisting of the crust and rigid mantle (see diagram, page 120).

(1) *Divergent Plate Boundary* – "The spreading boundary" – Along this boundary two plates are moving apart, forming a ridge where magma exits, creating new ocean floor. The Mid-Atlantic Ridge is a well-known and studied divergent plate boundary. At all ocean ridges, the newest rocks of the ocean floor are made as the molten magma surfaces and quickly solidifies into igneous basaltic rock. Moving away from these ocean ridges, the age of the ocean floor (ocean plate) increases.

(2) *Convergent Plate Boundary* – "The collision boundary" – Along this boundary two plates are moving toward each other. At a convergent boundary, the denser oceanic plate will dive or sink under the continental plate. This produces a subduction zone, making an ocean trench. The subducting oceanic plate will melt within the hot mantle, recycling this ocean floor into new magma. Some of this magma may reach the surface near this subducting zone, producing volcanoes.

When two continental plates collide, instead of subducting, they undergo uplifting (being less dense than ocean plates), producing large folded mountains that reach heights greater than 20,000 feet. The Himalayas, the highest mountain range in the world, was formed by this process.

(3) *Transform Plate Boundary* – "The slipping boundary"– Along this boundary, two plates are moving past each other. The most famous one is in California, known as the San Andreas Fault line. Here the North American Plate and the Pacific Plate are slipping by each other. The trouble associated with transform plate boundaries (and convergent boundaries) is that the plates become stuck, building up much pressure. Eventually, when the plates move, they release huge amounts of energy, causing major earthquakes, as Californians know all too well.

How were the plate boundaries discovered and mapped? The answer to this took years of research by many contributing geologists and scientists. A simple but efficient method was by plotting the location of hundreds of earthquakes. The boundary regions of plates are constantly grinding and moving, setting off numerous earthquakes. Volcanoes and mountain ranges are also located along plate boundaries.

The Map:

Key Area – On the bottom are located different symbols used on the map. The divergent plate boundary has opposite arrows showing the spreading action along mid-ocean ridges. The key for the convergent plate boundary shows which plate is the overriding plate (the less dense one) and which plate is the subducting plate (the denser one). For example, at the Aleutian Trench (by Alaska), the overriding plate is the northern part of the North American Plate and the subducting plate is the Pacific Plate. This subducting Pacific Plate produced the Aleutian Trench, a very deep part of the ocean floor. The transform plate boundary key shows the slippage of the plates by two arrows side-by-side going in opposite directions.

World Map – The major plate boundaries are outlined on the Tectonic Plates world map. The given arrows indicate the relative motions of the plates at their boundaries. Notice that the Pacific Plate is the largest plate. Along its boundary are found numerous active volcanoes, known as the "Ring of Fire". In the western Pacific Ocean is found the Mariana Trench. This subduction zone has produced the deepest trench of the world's oceans (6.85 mi). The Andes Mountains, located on the western coast of South America, were formed by the converging Nazca and South American Plates. Above India, locate the convergent boundary of the Indian-Australian Plate and the Eurasian Plate. Both of these convergent plates are continental plates. This collision produced the world's highest mountain, Mount Everest.

Mantle Hot Spots – This map shows nine hot spots. A hot spot is a volcanically active area that often is not on a plate boundary. Hot spots remain in the same position generating magma and producing volcanoes, while the plate slowly drifts over this area. Over time, as the plate moves, the volcano will be displaced off the hot spot and become extinct; however, a new volcano will slowly be formed over the hot spot as magma rises from the mantle. This process developed the Hawaiian Island chain.

Additional Information:

- New oceanic floor gets created at mid-ocean ridges and then gets melted in subduction zones. This recycling event causes the oceanic plates to be younger than the continental plates.

- Iceland was formed on the Mid-Atlantic Ridge and is labeled a Mantle Hot Spot. Active volcanoes, earthquakes, and hot springs are found in Iceland.

Diagrams:

1. **Transform Plate Boundary** – The fault line represents the San Andreas fault. On the left of this transform plate boundary is the Pacific Plate, on the right is the North American Plate. Along this boundary the plates slip by in opposite directions. Major earthquakes take place when movement occurs – releasing much stored up energy. Notice how thin the oceanic crust is compared to the continental crust.

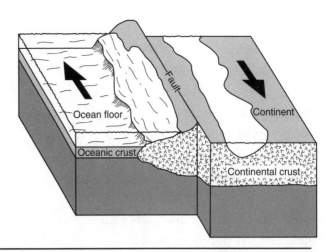

2.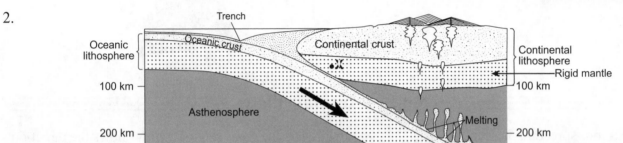

   Convergent Plate Boundary – When an oceanic plate converges with a continental plate, the denser oceanic plate subducts causing the ocean floor to depress forming an ocean trench. As the ocean plate dives deep into the hot mantle, it melts, and some of the magma rises to the surface, breaking through the crust, forming volcanoes.

3.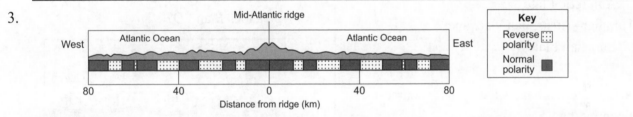

   Magnetic Polarity – The cross section shows a pattern of magnetic field (polarity) preserved in the igneous bedrock of the oceanic crust east and west of the Mid-Atlantic Ridge (MAR). As new ocean floor spreads from the MAR, the same polarity pattern is recorded east and west, at times being normal and at times being reversed. Also if the distances east and west from the MAR are the equal, the age of the igneous ocean floor will be the same, because they were formed at the same time at the MAR and then drifted in opposite directions. Magnetic polarity pattern and the age of the ocean floor are strong proofs of sea-floor spreading due to plate tectonics.

4. **Divergent Plate Boundary, Iceland** – Iceland is situated on the Mid-Atlantic Ridge – a divergent plate boundary. It was formed from a mantle hot spot producing volcanoes, lava flows, hot springs, and earthquakes, as magma rises to the surface. The plates on both sides of this island drift in opposite directions. The youngest rocks are located on the MAR and the rocks are older east and west of this ridge.

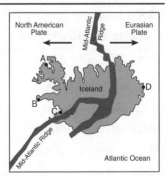

5. **Formation of a Rift Valley** – This diagram shows details of a rift valley. Here the crust thins, creating faults and fractures (shown by the vertical lines in the diagram) and eventually allowing magma to reach the surface producing volcanic activity. At a rift valley, a new divergent plate boundary is forming, and the continuation of this process will produce new ocean floor and the development of an ocean basin by the action of sea-floor spreading.

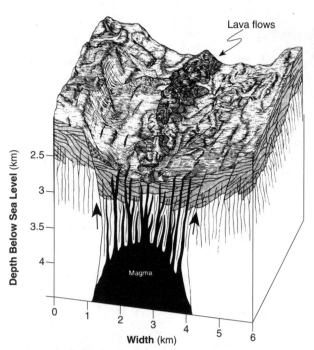

Set 1 — Tectonic Plates

1. Sea-floor spreading is occurring at the boundary between the

   (1) North American Plate and the Pacific Plate
   (2) Nazca Plate and South American Plate
   (3) Indian–Australian Plate and the Antarctic Plate
   (4) Indian–Australian Plate and Eurasian Plate   1_____

2. Earth's internal heat is the primary source of energy that

   (1) warms the lower troposphere
   (2) melts glacial ice at lower altitudes
   (3) moves the lithospheric plates
   (4) pollutes deep groundwater with radioactivity   2_____

3. Which feature is commonly formed at a plate boundary where oceanic crust converges with continental crust?

   (1) a mid-ocean ridge
   (2) an ocean trench
   (3) a transform fault
   (4) new oceanic crust   3_____

4. The cross section below shows the direction of movement of an oceanic plate over a mantle hot spot, resulting in the formation of a chain of volcanoes labeled A, B, C, and D. The geologic age of volcano C is shown. What are the most likely geologic ages of volcanoes B and D?

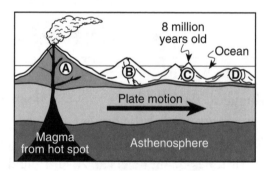

   (1) B is 5 million years old and D is 12 million years old.
   (2) B is 2 million years old and D is 6 million years old.
   (3) B is 9 million years old and D is 9 million years old.
   (4) B is 10 million years old and D is 4 million years old.   4_____

5. Which cross section below best represents the crustal plate motion that is the primary cause of the volcanoes and deep rift valleys found at mid-ocean ridges?

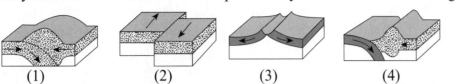

Key	
▨	Continental crust
▦	Oceanic crust
☐	Mantle
→	Direction of plate motion

(1)   (2)   (3)   (4)

5 _____

6. The accompanying diagram shows some features of Earth's crust and upper mantle.

Which model most accurately shows the movements (arrows) associated with the surface features shown in the diagram?

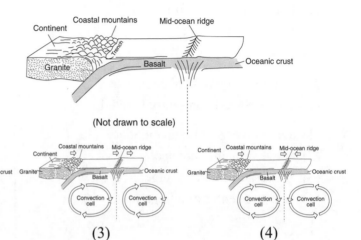

(Not drawn to scale)

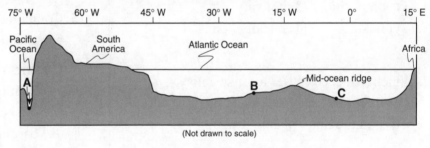

(1)   (2)   (3)   (4)

6 _____

The cross section below shows the major surface features of Earth along 25° S latitude between 75° W and 15° E longitude. Points *A*, *B*, and *C* represent locations on Earth's crust.

7. *a)* Identify the crustal feature located at point *A*.

_____

b) Identify the tectonic plate motion that is causing an increase in the distance between South America and Africa. _____

c) Bedrock samples were taken at the mid-ocean ridge and points *B* and *C*. On the grid, draw a line to show the relative age of the bedrock samples between these locations.

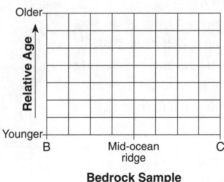

8. Identify the tectonic feature responsible for the formation of the Hawaiian Islands.

_____

9. Beneath which surface location is Earth's crust the thinnest?

   (1) East Pacific Ridge
   (2) the center of South America
   (3) Old Forge, New York
   (4) San Andreas Fault   9 _____

10. According to tectonic plate maps, New York State is presently located

    (1) at a convergent plate boundary
    (2) above a mantle hot spot
    (3) above a mid-ocean ridge
    (4) near the center of a large plate   10 _____

11. The observed difference in density between continental crust and oceanic crust is most likely due to differences in their

    (1) composition (3) porosity
    (2) thickness   (4) rate of cooling 11 _____

12. Which evidence causes most scientists to believe that sea-floor spreading occurs at the mid-Atlantic Ridge?

    (1) Oceanic crust is oldest at the ridge.
    (2) Large sedimentary folds exist in the mantle near the ridge.
    (3) Oceanic crust on both sides of the ridge is less dense than continental crust.
    (4) Oceanic crust on both sides of the ridge shows matching patterns of reversed and normal magnetic polarity. 12 _____

13. In which Earth layer are most convection currents that cause sea-floor spreading thought to be located?

    (1) crust   (3) outer core
    (2) asthenosphere   (4) inner core 13 _____

14. Which diagram best shows the type of plate boundary found between the Eurasian Plate and the Philippine Plate?

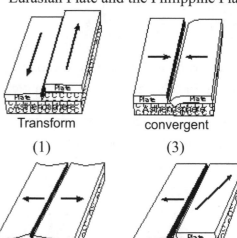

Transform   convergent
   (1)   (3)

Divergent   Complex
   (2)   (4)   14 _____

15. Compared to the oceanic crust, the continental crust is usually

    (1) thicker, with a less dense granitic composition
    (2) thicker, with a more dense basaltic composition
    (3) thinner, with a less dense granitic composition
    (4) thinner, with a more dense basaltic composition   15 _____

16. Hot spots are located

    (1) only at mid-ocean ridges
    (2) only on the continental plates
    (3) only on the oceanic plates
    (4) on both continental and ocean plates   16 _____

17. The diagrams below show four major types of fault motion occurring in Earth's crust. Which type of fault motion best matches the general pattern of crustal movement at California's San Andreas Fault?

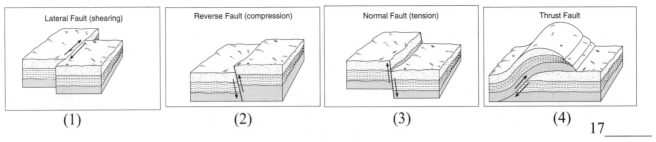

Lateral Fault (shearing)	Reverse Fault (compression)	Normal Fault (tension)	Thrust Fault
(1)	(2)	(3)	(4)

17_____

18. Which map best indicates the probable locations of continents 100 million years from now if tectonic plate movement continues at its present rate and direction?

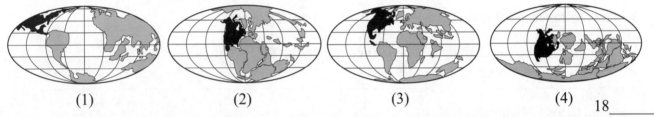

(1)   (2)   (3)   (4)

18_____

19. The accompanying diagram represents the pattern of normal and reversed magnetic polarity and the relative age of the igneous bedrock composing the ocean floor on the east side of the Mid-Atlantic Ridge. The magnetic polarity of the bedrock on the west side of the ridge has been deliberately left blank.

Which diagram best shows the magnetic pattern and relative age of the igneous bedrock on the west side of the ridge?

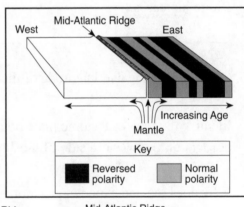

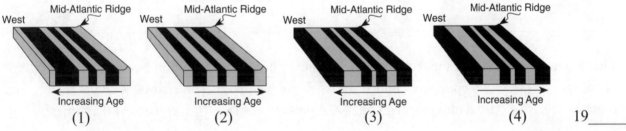

(1)   (2)   (3)   (4)   19_____

20. The Galapagos Hot Spot is located closest to what type of tectonic plate boundary?_____

21. Identify a process occurring in the plastic mantle that is inferred to cause tectonic plate motion.

_____

22. Identify the type of tectonic plate movement that caused Africa to separate from South America.

_____

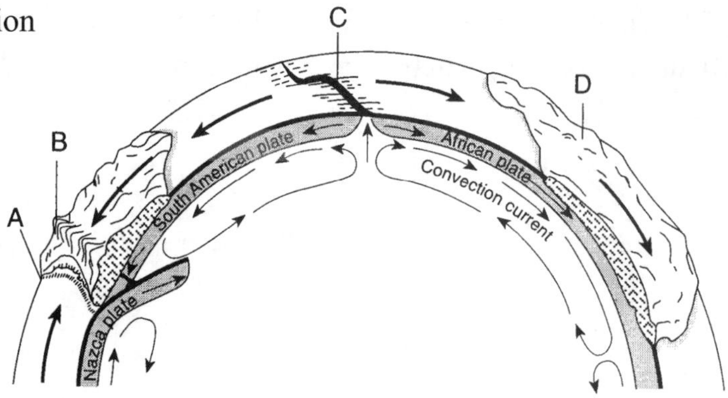

The diagram shows a cross section of a portion of Earth. The inferred motions of crustal plates are shown. Letters *A* through *D* represent locations at Earth's surface.

23. *a*) Which letter represents the location of the mid-Atlantic Ridge?

    (1) *A* (2) *B* (3) *C* (4) *D*

    a _____

b) The diagram shows convection currents. What are their roles with plate tectonics?

_____

c) The most recently formed ocean floor is located at which letter? _____

d) Explain why the Nazca plate is subducting beneath the South American plate.

_____

24. Explain why the age of the ocean-floor bedrock increases as the distance from the Mid-Atlantic Ridge increases.

_____

25. Give the latitude and longitude of the Canary Islands Hot Spot. _____N _____W

26. Identify the type of tectonic plate boundary that borders on the western side of the Juan de Fuca plate. _____

27. Why does Iceland experience major volcanic activities? _____

_____

28. The diagram below shows the tectonic plate boundary between Africa and North America 300 million years ago, as these two continents united into a single landmass. The arrows at letters *A*, *B*, *C*, and *D* represent relative crustal movements. Letter **X** shows the eruption of a volcano at that time.

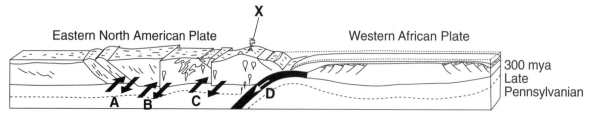

a) Identify the type of tectonic plate motion represented by the arrow shown at *D*. _____

b) Identify the type of tectonic motion represented by the arrows shown at *A*, *B*, and *C*. _____

Base your answers to question 29 on the diagram below, which shows an incomplete concept map identifying the types of plate boundaries. Information in the boxes labeled *A*, *B*, *C*, *D*, and *E* has been deliberately omitted.

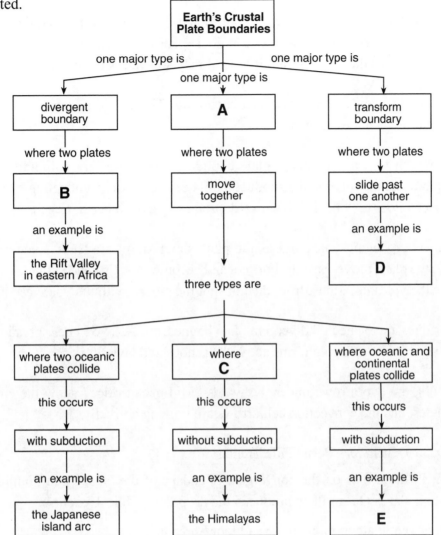

29. On the chart below, write the information that should be placed in the boxes labeled *A*, *B*, *C*, *D*, and *E* that will correctly complete those portions of the concept map.

Letter	Information That Should Be Placed in Each Box
A	
B	
C	
D	
E	

Tectonic Plates
Set 1 – Answers

1. 3 The key shows that, at the intersection of the Indian-Australian and Antarctic Plates, a divergent plate boundary exists. Here the newest ocean floor (plate) is made as escaping magma solidifies at the ridges. This forces the older ocean floor to slowly spread apart, producing sea-floor spreading.

2. 3 The heat within the mantle sets up convection currents. These flowing currents, located under the lithosphere, cause the plates to drift.

3. 2 The key shows that, at a convergent plate boundary, two plates are colliding, where one is the overriding plate and the other is the subducting plate. The denser, subducting oceanic crust (plate) causes the ocean floor to "buckle" downward, producing a deep ocean trench.

4. 1 The hot spot is producing volcanoes as the plate slowly moves over it. Eventually the volcano, connected to the plate, moves past the hot spot and becomes inactive and extinct. These sequences of events make the volcano farthest from the hot spot the oldest, with the youngest over the hot spot.

5. 3 Mid-ocean ridges are located on divergent plate boundaries. As two plates spread apart, rift valleys develop, while escaping magma forms active volcanoes and lava flows.

6. 1 In diagram (1), the two arrows under the words "Mid-ocean ridge" show the correct motion for divergent plates, and the convection cell arrows are flowing correctly.

7. *a*) Answer: Ocean Trench *or* Peru-Chile Trench

    Explanation: Position *A* is on the convergent boundary of the Nazca and South American Plate. The subducting Nazca Ocean Plate produced the deep Peru-Chile Trench.

    b) Answer: Divergent Plate motion *or* Sea Floor Spreading

    Explanation: At the mid-ocean ridge, two plates are diverging or spreading apart. Here lava forms the newest ocean floor (plate) causing the older ocean floor to slowly move away from the ridge.

    c)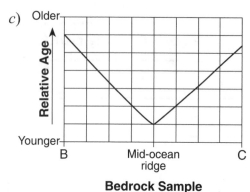

    Explanation: Point *B* and *C* are equally distant from the mid-ocean ridge, making them close to the same age. Years ago, the bedrock at points *B* and *C* was produced at the same time, at the ridge, as new magma solidified. Due to the motion at a divergent plate boundary, the spreading ocean floor caused point *B* and point *C* to move away from each other at nearly the same rate to their present locations.

8. Answer: Hot spot   Explanation: The chart shows that Hawaii sits on a mid-ocean hot spot.

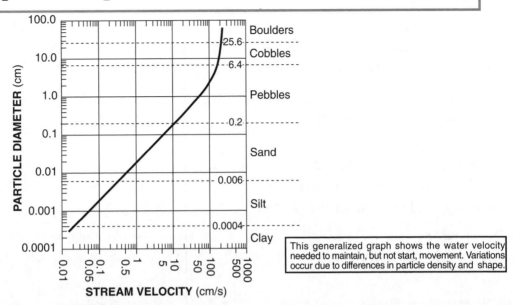

This generalized graph shows the water velocity needed to maintain, but not start, movement. Variations occur due to differences in particle density and shape.

Overview:

Exposed bedrock is reduced to sediments by the action of weathering. These smaller sediments can now be transported by water, wind, ice, and gravity. Of all of these agents of erosion, running water is the most powerful. As the water flows, the smallest sediments are first to be picked up. As the velocity (speed) of the water increases, the larger sediments begin to move. When the water loses energy and slows down, by entering a quiet body of water or by a decrease in the gradient, these sediments undergo deposition. Now the reverse occurs, the larger sediments are deposited (released) first, while the smaller sediments are the last to be deposited by the slower moving water.

The Graph:

The x-axis represents the velocity of water measured in cm/s. The y-axis represents the particle diameter in cm. The size range for each different sediment type is located on the right side of the chart. The range shows the lower and the upper limit for each sediment type. For example, using silt, the smallest possible size for this this sediment is 0.0004 cm, its largest size is 0.006 cm.

The graph line identifies the type of sediment that can be transported at a specific velocity. As expected, as the velocity increases, the graph line moves upward to the right, showing the larger particles that can be transported by faster water. When given a certain speed, move up from its position on the x-axis until the intersection of the graph line. This intersection position identifies the largest sediment that the water can carry. For example, at 100 cm/s the intersection point on the graph line is in the Pebbles section. At this speed, water will be able to transport medium size pebbles as well as all other smaller sediments. Sediments larger than pebbles will not be transported at this speed. If the water velocity decreases, deposition will occur. For example, if the velocity decreases from 100 cm/s to 1 cm/s, pebbles will be deposited first, then larger sand particles. This is how a sorted pattern of sediments develops in a water deposition system. Notice that clay, being the smallest particle, will be the first to be carried away by moving water, and this size sediment is the last to be deposited, needing a quiet body of water to do so.

Additional Information:

- Abrasion is the wearing away of a rock as it is transported. This occurs in streams as sediments bounce off each other. The end results are smaller, rounder sediments.

1. What is the largest particle that can be kept in motion by a stream that has a velocity of 100 centimeters per second?

   (1) silt   (3) pebble
   (2) sand   (4) cobble   1 _____

2. What is the largest particle that can generally be transported by a stream moving at 200 centimeters per second?

   (1) boulder   (3) pebble
   (2) cobble   (4) sand   2 _____

3. What is the minimum rate of flow at which a stream of water can maintain the transportation of pebbles 1.0 centimeter in diameter?

   (1) 50 cm/s   (3) 150 cm/s
   (2) 100 cm/s   (4) 200 cm/s   3 _____

4. A stream velocity decreases from 5 to 0.5 centimeters per second. Which statement would be true?

   (1) Clay, silt, sand, pebbles, and smaller cobbles stay in transport; some cobbles are deposited.
   (2) Clay, silt, sand, and smaller pebbles stay in transport; some pebbles are deposited.
   (3) Clay, silt, and smaller sand stay in transport; some sand is deposited.
   (4) Clay and smaller silt stay in transport; some silt is deposited.   4 _____

5. When a stream's velocity decreases from 300 to 200 centimeters per second, which size sediment will be deposited?

   (1) boulders   (3) sand
   (2) pebbles   (4) silt   5 _____

Note: Question 6 has only three choices.

6. As the velocity of a stream decreases, the amount of sediment being transported by the water of the stream.

   (1) decreases
   (2) increases
   (3) remains the same   6 _____

7. The cross section below illustrates the general sorting of sediment by a river as it flows from a mountain to a plain.

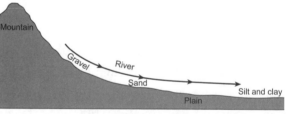

(Not drawn to scale)

   Which factor most likely caused the sediment to be sorted in the pattern shown?
   (1) velocity of the river water
   (2) hardness of the surface bedrock
   (3) mineral composition of the sediment
   (4) temperature of the water   7 _____

8. Which particle of quartz shows evidence of being transported the farthest distance by the stream?

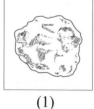  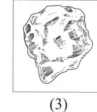 

    (1)   (2)   (3)   (4)   8 _____

9. A stream flowing at a velocity of 75 centimeters per second can transport

    (1) clay, only
    (2) pebbles, only
    (3) pebbles, sand, silt, and clay, only
    (4) boulders, cobbles, pebbles, sand, silt, and clay   9_____

10. What is the minimum water velocity necessary to maintain movement of 0.1-centimeter-diameter particles in a stream?

    (1) 0.02 cm/s   (3) 5.0 cm/s
    (2) 0.5 cm/s   (4) 20.0 cm/s   10_____

11. A stream with a velocity of 100 centimeters per second flows into a lake. Which sediment-size particles would the stream most likely deposit first as it enters the lake?

    (1) boulders   (3) pebbles
    (2) cobbles   (4) sand   11_____

12. Which size particle will remain suspended longest as a river enters the ocean?

    (1) pebble   (3) silt
    (2) sand   (4) clay   12_____

13. As water velocity of a stream increases from 0.2 to 200 centimeters per second, in which order will particles of different sizes begin to move?

    (1) sand → pebbles → cobbles → boulders   (3) cobbles → pebbles → sand → silt
    (2) silt → sand → pebbles → cobbles   (4) silt → pebbles → sand → cobbles   13_____

The accompanying profile shows the average diameter of sediment that was sorted and deposited in specific areas A, B, C, and D by a stream entering an ocean.

14. As compaction and cementation of these sediments eventually occur, which area will become siltstone?

    (1) A   (2) B   (3) C   (4) D   14_____

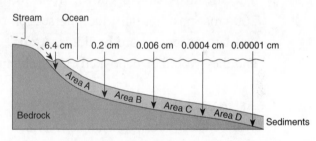

15. Which graph best shows the general relationship between stream velocity and the diameter of particles transported by a stream?

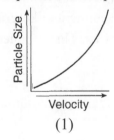
(1)

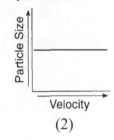
(2)

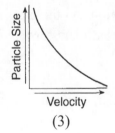
(3)

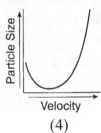
(4)

    15_____

16. Identify *two* factors that determine the rate of stream erosion.

1_____   2_____

Relationship of Transported Particle Size to Water Velocity

Answers

Set 1

1. 3 Open to the Relationship of Transported Particle Size to Water Velocity graph. On the Stream Velocity axis, locate the 100 cm/s position. Go directly up from this speed until the intersection of the graph line. This occurs in the Pebble section. This would be the largest sediment that can be transported by a stream that is moving at 100 cm/s.

2. 2 On the Stream Velocity axis, locate the 200 cm/s position. Go directly up from this speed until the intersection of the graph line. This occurs in the Cobbles section. This would be the largest sediment that can be transported by a stream that is moving at 200 cm/s.

3. 1 On the Particle Diameter axis, locate 1.0 cm. From this position, move directly to the right until the intersection of the graph line. This intersection point is in the Pebble area. From this position, move directly down to the Stream Velocity axis where the minimum rate of flow is given.

4. 3 The graph shows that at 5 cm/s, the largest particles the stream can transport are the large sand particles. When the stream's speed reduces to 0.5 cm/s, the stream can transport only the smallest sand particles, along with silt, and clay. Thus, larger sand particles will be deposited at this reduced speed.

5. 1 The stream is slowing down from 300 to 200 cm/s. From the Stream Velocity axis, go up the 300 cm/s position until the intersection of the graph line. At this speed, the largest sediment that can be transported are boulders. At 200 cm/s, the stream is no longer able to transport boulders, and they would be deposited on the channel floor.

6. 1 As the velocity decreases, deposition increases, and the stream would be transporting fewer sediments. If you follow the graph line from the Boulders at 300 cm/s to a slower speed (moving to the left), the graph line shows the larger sediments will be deposited first, then the smaller ones.

7. 1 The velocity or speed of a stream will decrease if the discharge (amount of water in a stream) decreases or the gradient decreases. The diagram shows that the gradient is high near the mountain and is almost level on the plain region. As the gradient decreases, the large sediments like gravel are deposited first, then the smaller sediments are released according to their sizes. This produces a sorted depositional pattern.

8. 4 As sediments are transported by water, abrasion occurs making the sediment smaller and rounder.

Rock Cycle in Earth's Crust

Overview:

All rocks can be classified as igneous, metamorphic, or sedimentary. As permanent as rocks appear, eventually geological forces act on them, causing major changes. These changes may alter the rock to be reclassified into a different family of rocks. Even rocks that have been melted within the mantle forming magma may one day surface as lava, solidifying only to be exposed to the forces of the "Rock Cycle." So the real question is: "Are rocks ever really destroyed?"

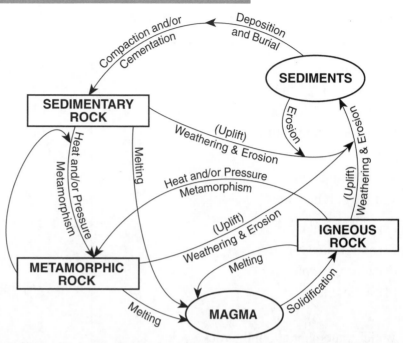

The Chart:

The outside circle shows the processes and steps that change rocks from one family into another. This path is not a one-way route. The rock cycle shows many other paths represented by inner circles. Following all the arrows representing the different paths, it would appear that any path is possible for a particular rock. This is almost true. Let's look at each type of rock family and how it fits in the rock cycle.

Igneous Rocks – The initial process for an igneous rock is melting, producing magma. As this magma/lava cools, it undergoes the process of solidification, changing the molten material to a solid igneous rock. The igneous rock may be remelted and cycled back to magma; or it may be subjected to heat and/or pressure, changing it to a metamorphic rock; or over time, weathering and erosion can break down the igneous rock into sediments for future sedimentary rocks.

Sedimentary Rocks – Sediments are smaller pieces of rocks that are easily moved by wind and water, and may eventually settle in a lake, shallow sea, or by the ocean shoreline. Locate Sediments. The path shows that these sediments are deposited and buried, almost always underwater, becoming compacted (compaction) from the weight above them. Dissolved minerals slowly cement (cementation) the sediments together, producing sedimentary rocks. If these rocks are uplifted, emerging out of the water, weathering forces will reduce them to smaller sediments, and erosion will transport them away to start another sedimentary cycle. But sedimentary rocks can take two other paths, the metamorphic rock path or the igneous rock path. Both paths involve different processes.

Metamorphic Rocks – Heat and/or pressure applied to rocks may cause them to change into metamorphic rocks. Notice all arrows heading to metamorphic rocks have this process of heat and/or pressure as a requirement. The arrows (paths) show all rocks can undergo metamorphism, including a metamorphic rock. Plate tectonics is the major force producing the heat and pressure of metamorphism.

Additional Information:

- When heat and pressure is added to a rock, the chart shows the resulting rock is metamorphic. If too much heat is added and melting occurs, upon solidification the resulting rock must be igneous.

1. Rocks are classified as igneous, sedimentary, or metamorphic based primarily on their

   (1) texture
   (2) crystal or grain size
   (3) method of formation
   (4) mineral composition   1_____

2. When granite melts and then solidifies, it becomes

   (1) a sedimentary rock
   (2) an igneous rock
   (3) a metamorphic rock
   (4) sediments   2_____

3. Which statement about the rock cycle is *not* true?

   (1) Cementation is a process that leads to sedimentary rocks.
   (2) When heat is applied to a rock and it melts, it may form a metamorphic rock.
   (3) A sedimentary rock in the future may change into another type of sedimentary rock.
   (4) Solidification is always needed to form an igneous rock.   3_____

4. The burial process involving sedimentary rocks is usually

   (1) deep within the Earth.
   (2) at the surface of the Earth.
   (3) within a water environment.
   (4) at an ocean ridge.   4_____

5. The solidification of magma produces

   (1) igneous and metamorphic rocks.
   (2) sedimentary rocks and igneous rocks.
   (3) only igneous rocks.
   (4) only metamorphic rocks.   5_____

6. Which statement about a metamorphic rock is not supported by the rock cycle?

   (1) A metamorphic rock may become sediments.
   (2) Metamorphic rocks may one day undergo melting.
   (3) A metamorphic rock has undergone cementation.
   (4) A metamorphic rock may eventually become another type of metamorphic rock.   6_____

The boxes labeled *A* through *G* represent rocks and rock materials. Arrows represent the processes of the rock cycle.

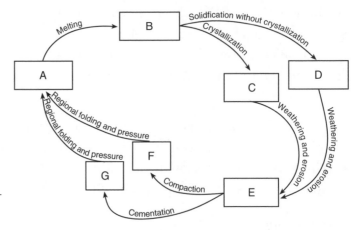

7. *a*) Which lettered box could represent a sedimentary rock?
   (1) *E*   (3) *C*
   (2) *G*   (4) *D*   7_____

b) From this diagram, give the process to change a preexisting rock to a metamorphic rock.

8. The process of uplift is essential to the rock cycle for

   (1) increasing the weight necessary for compaction.
   (2) forming ocean trenches where erosion is dominant.
   (3) exposing rocks to the forces of weathering and erosion producing sediments.
   (4) increase the temperature needed to produce magma.   8_____

9. The crystals of many metamorphic rocks are aligned in bands as a result of

   (1) earthquake faulting
   (2) cooling and solidification
   (3) mechanical weathering
   (4) heat and pressure   9_____

10. Which statement about an igneous rock is not supported by the rock cycle?

    (1) All igneous rocks eventually become sedimentary rocks.
    (2) An igneous rock may be remelted and solidified into another type of igneous rock.
    (3) All igneous rocks must have been a liquid at one time.
    (4) An igneous rock can undergo metamorphism.   10_____

11. Uplift, weathering and erosion leads to the formation of

    (1) magma
    (2) sediments
    (3) cementation
    (4) igneous rocks   11_____

12. State the rock family that should be listed in the rectangle for:

    Rock A _____

    Rock B _____

    Rock C _____

High-grade metamorphism in which some mica has changed to feldspar; banding is present

Compaction and cementation

Weathering and erosion to fragments less than 0.0004 cm in diameter

Rock A

Rock B

Rock C

Melting

Slow cooling underground

Felsic magma

13. Give the processes to form an igneous rock.

_____

14. Give the processes to form a sedimentary rock.

_____

Rock Cycle in Earth's Crust

Answers

Set 1

1. 3 The three rock families are based on how they were formed. The Rock Cycle in Earth's Crust chart shows that igneous rocks originate from magma, while sedimentary rocks are made by the compaction and/or cementation of sediments. Metamorphic rocks have been changed by the processes of heat and pressure.

2. 2 When any rock melts then solidifies, an igneous rock will always be formed.

3. 2 If heat is applied to any rock and the rock melts, magma is produced. The Rock Cycle chart shows that when magma cools and solidifies, it always produces igneous rocks.

4. 3 Sediments are easily transported by moving water. As water loses its energy, these sediments will be deposited (settled) and buried in some water environment such as a lake, shallow sea, or ocean.

5. 3 Under the right conditions, any type of rock can be melted changing it to magma. But once magma solidifies, it becomes an igneous rock. The Rock Cycle chart clearly shows this one-way path.

6. 3 Sediments that undergo compaction and/or cementation produce sedimentary rocks. Metamorphic rocks have undergone the process of heat and/or pressure.

7. *a*) 2 Sedimentary rocks are formed from sediments that have been cemented and/or compacted. These processes cause sediments to form sedimentary rocks. From the diagram, both letters *G* and *F* would produce a sedimentary rock.

   b) Answer: Regional folding and pressure *or* heat and pressure

   Explanation: Regional folding is associated with plate tectonics. As plates collide, much pressure and heat develops causing rock layers to fold. These conditions will cause metamorphism to a large segment of the preexisting rocks.

Scheme for Igneous Rock Identification

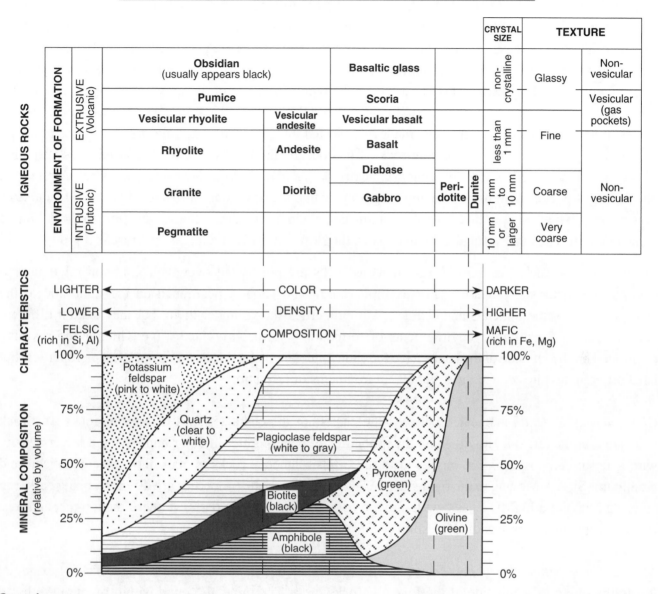

Overview:

Igneous rocks form as molten rock (lava/magma) cools and solidifies. If this solidification process occurs inside the Earth, the magma will form intrusive igneous rocks. When volcanic action brings magma to the surface, it quickly solidifies, creating an extrusive igneous rock. Both of these classifications exhibit different textures based on the size of the crystals that grow. The slower the magma cools, the larger the resulting crystals and the coarser the texture. Lava usually cools so fast that no visible crystals develop. This produces a glassy, non-crystalline rock.

All rocks are composed of minerals. Two igneous rocks may have the same minerals, but have different names. This occurs if they formed from different environments – intrusive or extrusive. The characteristic properties of density, color, and mineral composition, along with texture, are used to identify igneous rocks.

The Chart:

Igneous Rocks section – On the left side is given the Environment Of Formation – Extrusive (Volcanic), meaning outside the Earth, and Intrusive (Plutonic), meaning inside the Earth. Igneous rocks are first classified into one of these environments in which different textures are produced. In the Texture columns are given the different textures for extrusive and intrusive igneous rocks. In extrusive environments, glassy texture may be present. This is caused by the rapid solidification of lava, producing no observable crystals. This is referred to as non-crystalline (see Crystal Size column). Fine texture is when the crystals are less than 1 mm in size. At times, magma/lava is ejected from a volcano and solidifies in the air, producing air holes or gas pockets in the volcanic rock. If this occurs, the rock's texture is referred to as vesicular. This texture is an excellent clue for identifying extrusive igneous rocks. A non-vesicular extrusive rock will not show gas pockets, but it might exhibit a glassy or fine texture. In an intrusive environment, coarse or very coarse textures are produced. As the magma cools within the Earth, atoms have time to "lock" into a crystal pattern. The more insulated the magma is, the slower it cools and larger the crystals it grows.

In this Igneous Rocks section different igneous rocks are positioned over their common minerals and common characteristic properties. For example, basalt and gabbro, positioned on the right side of the chart, have the same minerals and characteristic properties shown under them, but they exhibit different textures due to the different environments of formation – intrusive vs. extrusive. Granite and rhyolite, positioned on the left side of the chart, have the same mineral composition and characteristic properties, but have different textures due to the different environments of formation.

Characteristics section – Located between the charts are the characteristics of igneous rocks. If a rock is located on the left side, it will show the properties of being lighter in color and having a lower density with a felsic composition. The chart shows these felsic igneous rocks contain the elements silicon and aluminum (Si, Al). Moving to the right side, the given characteristic properties are different, being darker in color and having a higher density with a mafic composition. The chart shows these mafic igneous rocks contain the elements iron and magnesium (Fe, Mg).

Mineral Composition chart – This chart gives the minerals found in the specific igneous rocks. For example, locate rhyolite on the left side of the Igneous Rocks chart. The minerals found within rhyolite are listed directly below in the Mineral Composition chart. These minerals would include potassium feldspar, quartz, plagioclase feldspar, and smaller amounts of biotite and amphibole. This chart also shows that the percentage of each mineral can vary. The same procedure is used to find the mineral composition of the other given igneous rocks. When this procedure is used, it becomes evident that the igneous rocks located on the left side of the chart, when compared to igneous rocks located on the right side of the chart, will have different mineral compositions.

Observing an unknown igneous rock's characteristic properties, and its texture, as well as identifying some of the minerals within, greatly helps in the identification of the rock.

Additional Information:

- Obsidian usually appears black, but when sliced into thinner sections, it is light in color and translucent.

Scheme for Igneous Rock Identification

1. **Igneous Environments** – A magma chamber is an underground pool of liquid rock. If the outer part of this chamber (A) cools enough to solidify, the resulting intrusive igneous rocks will have a coarse texture exhibiting large interlocking crystals. If the magma surfaces, the lava (B) will cool quickly, producing an extrusive igneous rock with a fine, glassy, or vesicular texture.

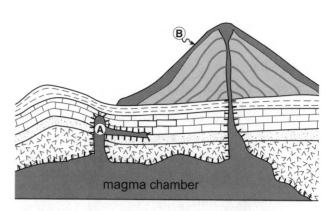

magma chamber

2.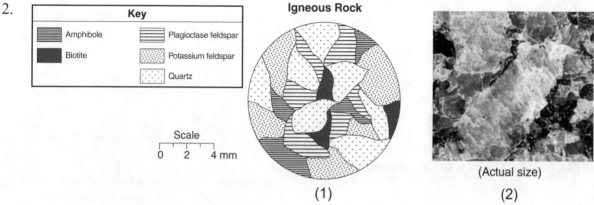

Igneous Rock

Key	
▤ Amphibole	▤ Plagioclase feldspar
■ Biotite	▨ Potassium feldspar
⋮ Quartz	

Scale
0   2   4 mm

(1)

(Actual size)

(2)

Intrusive Igneous Rock – Diagram 1 shows the magnified drawing of the minerals found in an intrusive igneous rock. Based on the size of its intergrown crystals, the environment of formation must have been intrusive (plutonic). Diagram 2 is an actual photograph of such an intrusive igneous rock that cooled slowly, growing large crystals.

3.

Mineral Name	Percentage of Mineral Present
plagioclase feldspar	55%
biotite	15%
amphibole	30%

0   1
centimeter

Identifying an Igneous Rock – The three given minerals and their mineral percentages would be positioned in the center section of the Mineral Composition chart. The name of this rock would be directly above this middle section within the Igneous Rock chart. Measurement of the grain sizes indicates that the texture is coarse. Diorite is situated above the given minerals and has a coarse texture.

4. **Extrusive Igneous Rocks** – Rock *A* has gas pockets, making its texture vesicular. This rock cooled rapidly, trapping air as lava was ejected from a volcano. This rock is vesicular basalt.

Rock *B* shows a glassy texture making it non-crystalline. A lava flow on the Earth's surface cooled quickly, producing this texture. This rock is obsidian.

(A)

vesicular texture

(B)

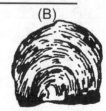

Glassy black rock that breaks with a shell-shape fracture

1. Which three minerals are most commonly found in the igneous rock granite?

    (1) amphibole, calcite, and hematite
    (2) amphibole, biotite mica, and gypsum
    (3) plagioclase feldspar, pyroxene, and olivine
    (4) plagioclase feldspar, potassium, feldspar, and quartz

    1_____

2. The three statements below are observations of the same rock sample:

    • The rock has intergrown crystals from 2 to 3 millimeters in diameter.

    • The minerals in the rock are gray feldspar, green olivine, green pyroxene, and black amphibole.

    • There are no visible gas pockets in the rock.

    This rock sample is most likely

    (1) sandstone   (3) granite
    (2) gabbro   (4) rhyolite

    2_____

3. Which igneous rock has a vesicular texture and a felsic composition?

    (1) pumice   (3) granite
    (2) basalt   (4) scoria

    3_____

4. Name an igneous rock with mineral crystals ranging in size from 2 to 6 millimeters. The rock is composed of 58% plagioclase feldspar, 26% amphibole, and 16% biotite. What is the name of this rock?

    (1) diorite   (3) andesite
    (2) gabbro   (4) pumice

    4_____

5. Which extrusive igneous rock could be composed of approximately 60% pyroxene, 25% plagioclase feldspar, 10% olivine, and 5% amphibole?

    (1) granite   (3) gabbro
    (2) rhyolite   (4) basalt

    5_____

6. Which rock most probably formed directly from lava cooling quickly at Earth's surface?

    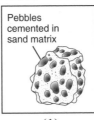
    Pebbles cemented in sand matrix
    (1)

    Mica crystals in foliated layers
    (3)

    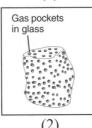
    Gas pockets in glass
    (2)

    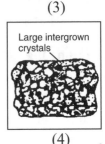
    Large intergrown crystals
    (4)

    6_____

7. Which characteristic provides the best evidence that obsidian rock formed in an extrusive environment?

    (1) layers of rounded fragments
    (2) distorted bands of large mineral crystals
    (3) noncrystalline glassy texture
    (4) mineral cement between grains

    7_____

8. For an igneous rock to be classified as rhyolite, it must be light colored, be fine grained, and contain

    (1) quartz   (3) pyroxene
    (2) calcite   (4) olivine

    8_____

9. The accompanying graph shows the relationship between the cooling time of magma and the size of the crystals produced.

Which graph correctly shows the relative positions of the igneous rocks granite, rhyolite, and pumice?

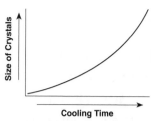

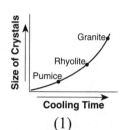

(1)

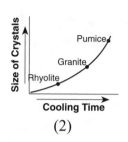

(2)

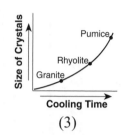

(3)

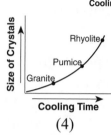

(4)

9_____

Base your answers to question 10 on the diagram and table below. The diagram represents a felsic igneous rock. Letters *A*, *B*, and *C* represent three different minerals in the rock sample. The table describes the physical properties of minerals *A*, *B*, and *C* found in the igneous rock sample.

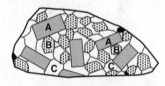

(Actual size)

Mineral	Key	Physical Properties
A		pink, cleaves in two directions at 90°
B		white, cleaves in two directions, striations visible
C		colorless or clear with a glassy luster

10. *a)* State the texture of this igneous rock. _____

b) State *two* processes responsible for the formation of an igneous rock.

1) _____ 2)_____

c) Using the Properties of Common Minerals chart (see page 207), give the name of:

Mineral *A* _____

Mineral *B* _____

Mineral *C* _____

11. What is the difference between the composition of felsic and mafic igneous rocks?

_____

12. Describe how a vesicular texture is produced in a volcanic rock.

_____

13. The end product of the weathering of gabbro or basalt rocks is a solution of dissolved material that most likely would contain high amounts of

    (1) iron and magnesium
    (2) magnesium and potassium
    (3) aluminum and iron
    (4) aluminum and potassium   13_____

14. The photograph below shows an igneous rock.

    What is the origin and rate of formation of this rock?

    (1) plutonic with slow cooling
    (2) plutonic with rapid cooling
    (3) volcanic with slow cooling
    (4) volcanic with rapid cooling   14_____

15. An igneous rock is a dark-colored crystalline rock that formed when a lava flow cooled and solidified quickly on the surface of Earth. This igneous rock is classified as an

    (1) extrusive igneous rock with a coarse texture and felsic composition
    (2) extrusive igneous rock with a fine texture and a mafic composition
    (3) intrusive igneous rock with a coarse texture and a felsic composition
    (4) intrusive igneous rock with a fine texture and a mafic composition
                                        15_____

16. Which igneous rock has a vesicular texture and contains the minerals potassium feldspar and quartz?

    (1) andesite   (3) pumice
    (2) pegmatite   (4) scoria   16_____

17. Which diagram represents a landscape where fine-grained igneous bedrock is most likely to be found?

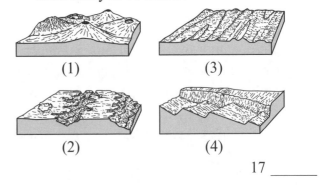

    (1)   (3)

    (2)   (4)
                                        17_____

18. Which diagram best represents a sample of an igneous rock?

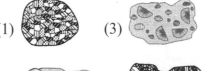

    (1)   (3)

    (2)   (4)   18_____

19. The photograph below shows actual crystal sizes in a light-colored igneous rock that contains several minerals, including potassium feldspar, quartz, and biotite mica.

    (Shown to actual size)

    The rock should be identified as

    (1) granite   (3) basalt
    (2) gabbro   (4) rhyolite   19_____

Base your answers to question 20 on the Rock Classification Flowchart. Letters *A*, *B*, and *C* represent specific rocks in this classification scheme.

Rock Classification Flowchart

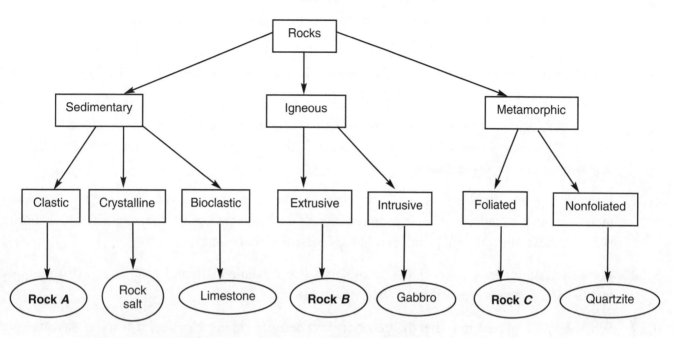

20. *a*) Rock *B* has a fine vesicular texture and is composed mainly of potassium feldspar and quartz. State the name of rock *B*. _____

b) Granite could be placed in the same position in the flowchart above as gabbro. Describe *two* differences between granite and gabbro.

1) _____

2) _____

c) If Rock *B* is scoria, give two descriptions of its properties. _____

_____

21. Complete the table below, with descriptions of the observable characteristics used to identify basalt.

Characteristic of Basalt	Description
Texture	
Color	
Density	

22. Name a nonvesicular rock made entirely of green 2-millimeter-diameter crystals that has a hardness of 6.5 and exhibits fracture (see page 207).

_____

Scheme for Igneous Rock Identification
Set 1 – Answers

1. 4 The minerals that make up an igneous rock are located directly under that rock in the Mineral Composition chart. The listed minerals under granite match those in choice 4.

2. 2 The listed minerals are located on the right side of the Mineral Composition chart. The sample rock would be positioned directly above these minerals in the Igneous Rock chart. Grain size 2-3 mm, would be coarse. Gabbro is the sample rock.

3. 1 Locate the vesicular (gas pockets) texture section. Moving to the left, pumice is aligned with this texture. As shown in the Characteristics section, felsic composition rocks are located on the left side, where pumice is positioned.

4. 1 Due to the grain size, this rock would be located in the coarse texture row. The listed minerals and their percentages are located in the center section of the Mineral Composition chart. The rock diorite is above these minerals and positioned in the coarse texture column.

5. 4 Locate basalt. Directly under it are the minerals that it would contain. Basalt is an extrusive rock that has the mineral composition given in the question.

6. 2 When lava is ejected into the air, gases are trapped producing a vesicular texture. Because the lava solidified quickly, a glassy texture also is exhibited.

7. 3 To the right of obsidian the crystal size and texture is given as non-crystalline and glassy. Extrusive igneous rocks cool very fast producing a smooth glassy texture.

8. 1 Open to the Igneous Rock chart and locate rhyolite. Look under rhyolite in the Mineral Composition chart. It shows that rhyolite contains quartz in its composition.

9. 1 Granite, an intrusive rock having the largest crystals, must have had the longest cooling time. Pumice, with a glassy texture, producing no visible crystals, must have cooled the fastest of the rocks.

10. *a*) Answer: Coarse

    Explanation: Coarse texture, having crystal sizes of 1 mm to 10 mm, is the correct range size of the minerals in the diagram.

    b) Answer: Melting and Solidification

    Explanation: Open to the Rock Cycle chart, page 79. The paths to Igneous Rock involve the processes of melting and solidification.

    c) Mineral *A* – Potassium Feldspar   Mineral *B* – Plagioclase Feldspar   Mineral *C* – Quartz

    Explanation: Use the given Physical Properties to identify the minerals using the Properties of Common Minerals chart (see page 207).

11. Felsic rocks contain silicon (Si) and aluminum (Al), while mafic rocks contain iron (Fe) and magnesium (Mg). See the Composition area in the Characteristics section.

12. A vesicular texture is where visible gas pockets are present. This occurs when lava solidifies in the air when ejected from a volcanic eruption.

Scheme for Sedimentary Rock Identification

INORGANIC LAND-DERIVED SEDIMENTARY ROCKS

TEXTURE	GRAIN SIZE	COMPOSITION	COMMENTS	ROCK NAME	MAP SYMBOL
Clastic (fragmental)	Pebbles, cobbles, and/or boulders embedded in sand, silt, and/or clay	Mostly quartz, feldspar, and clay minerals; may contain fragments of other rocks and minerals	Rounded fragments	Conglomerate	
			Angular fragments	Breccia	
	Sand (0.006 to 0.2 cm)		Fine to coarse	Sandstone	
	Silt (0.0004 to 0.006 cm)		Very fine grain	Siltstone	
	Clay (less than 0.0004 cm)		Compact; may split easily	Shale	

CHEMICALLY AND/OR ORGANICALLY FORMED SEDIMENTARY ROCKS

TEXTURE	GRAIN SIZE	COMPOSITION	COMMENTS	ROCK NAME	MAP SYMBOL
Crystalline	Fine to coarse crystals	Halite	Crystals from chemical precipitates and evaporites	Rock salt	
		Gypsum		Rock gypsum	
		Dolomite		Dolostone	
Crystalline or bioclastic	Microscopic to very coarse	Calcite	Precipitates of biologic origin or cemented shell fragments	Limestone	
Bioclastic		Carbon	Compacted plant remains	Bituminous coal	

Overview:

The origin of sedimentary rocks is almost always associated with a water environment. In lakes, shallow seas, and ocean shorelines, sediments are deposited, buried, compacted, and/or cemented, producing sedimentary rocks. These processes are shown on the Rock Cycle chart. Sedimentary rocks have a wide variation of sediment sizes, texture, and composition. To help organize these rocks, they are classified into two groups: *Inorganic Land-Derived Sedimentary Rocks* and *Chemically and/or Organically Formed Sedimentary Rocks*. Using this classification system along with the information within the chart, the 10 given rocks can be successfully identified.

The Chart:

Inorganic Land-Derived Sedimentary Rocks – This upper section consists of 5 inorganic (non-living) rocks (see Rock Name column). The origin of the sediments for these rocks were land-derived and were eventually compacted and/or cemented under water. These rocks have a clastic texture, meaning fragmented particles (sediments) make up the rock. These fragmented sediments might be microscopic clay particles or visible sand particles found in sandstone, or pebble size particles cemented in a conglomerate. These five rocks are organized by grain size, which is their sediment size. The smallest sediment is clay, being smaller than 0.0004 cm. When clay undergoes compaction and/or cementation, the resulting rock is shale. The largest grain sizes (sediments) are found in conglomerates. Because sedimentary rocks are a mixture of different rocks, their composition varies greatly. The Map Symbol column shows the designated diagrammed symbols for the given rocks.

Note: The lower section on the Sedimentary Rock chart, shown on the previous page, has been separated into two sections for clarity.

Chemically Formed Sedimentary Rocks – The chart shows three chemically formed sedimentary rocks: rock salt, rock gypsum, and dolostone, having crystalline texture (a texture having crystals). Each of these rocks has different mineral compositions, but all are formed by the evaporation of water. As water evaporates, the dissolved minerals become concentrated and start to precipitate (release) out of the water, settling to the bottom and building an evaporite sedimentary rock. This is how the salt layers in NYS were produced. In the Comments column are the terms "precipitates and evaporites."

Organically Formed Sedimentary Rocks – These rocks were formed from once living material, making a bioclastic or crystalline texture. Coal, having a bioclastic texture, is composed of carbon from the compaction of trees and plant remains. The other given example of a bioclastic or crystalline texture is limestone. The Comments section for limestone states "Precipitates of biologic origin or cemented shell fragments." Limestone contains the mineral calcite, which reacts by bubbling when in contact with an acid. This is why an acid test is useful in identifying limestone.

Additional Information:

- Breccia is a type of conglomerate. The difference between breccia and a conglomerate is that breccia shows angular fragments, while a conglomerate has mostly rounded sediments.

Diagrams:

1. **Sedimentary Rocks and Fossils** – Almost all sedimentary rocks were formed under water. When marine organisms die, some are preserved in the marine sediments and fossilized as the sediments become cemented and/or compacted, changing into a sedimentary rock. A great clue that a rock is sedimentary is the presence of fossils – like this Paleozoic age starfish.

0   2 cm

2. **Sediments to Rocks** – This diagram represents the deposition of sediments when a river enters a lake. Letter *C* is where the river entered the lake depositing the larger and heavier pebbles. Letter *A* is farther out in the lake where the smaller clay particles settled. Over years and with compaction and/or cementation, these sediments of different sizes produce different sedimentary rocks. Clay becomes shale, silt becomes siltstone, sand becomes sandstone and a mixture of pebbles and smaller sediments produce the sedimentary rock known as conglomerate.

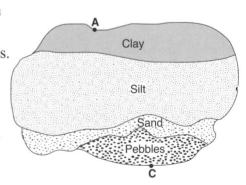

3. **Identification of a Sedimentary Rock** – The diagram represents a sedimentary rock composed of cemented pebbles and sand. From the Sedimentary Rock chart, this rock would be classified as an inorganic land-derived with clastic (fragmental) texture. As detailed in the chart, this rock is a conglomerate because it contains pebbles and other size sediments. Remember, inorganic land-derived sedimentary rocks are based on grain size, not composition.

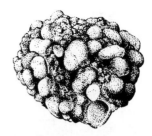

(Shown actual size)

4.

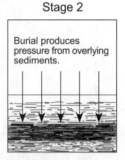 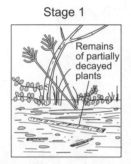 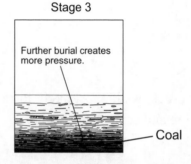

Stage 1 | Stage 2 | Stage 3

Remains of partially decayed plants

Burial produces pressure from overlying sediments.

Further burial creates more pressure.

Coal

Formation of Coal – Coal is classified as an organically formed sedimentary rock. The diagram shows the different stages of how decaying organic matter gets buried and undergoes compaction by the pressure of the overlying rocks. This process over time convert organic matter into bituminous coal.

Set 1 — Scheme for Sedimentary Rock Identification

Base your answers to question 1 on the drawings of six sedimentary rocks labeled *A* through *F*.

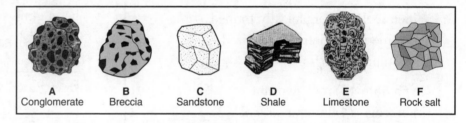

| A Conglomerate | B Breccia | C Sandstone | D Shale | E Limestone | F Rock salt |

1. *a)* Most of the rocks shown were formed by

(1) volcanic eruptions and crystallization
(2) compaction and/or cementation
(3) heat and pressure
(4) melting and/or solidification   a_____

b) Which two rocks are composed primarily of quartz, feldspar, and clay minerals?

(1) rock salt and conglomerate
(2) rock salt and breccia
(3) sandstone and shale
(4) sandstone and limestone   b_____

c) Which table shows the rocks correctly classified by texture?

(1)
Texture	clastic	bioclastic	crystalline
Rock	A, B, C, D	E	F

(2)
Texture	clastic	bioclastic	crystalline
Rock	A, B, C	D	E, F

(3)
Texture	clastic	bioclastic	crystalline
Rock	A, C	B, E	D, F

(4)
Texture	clastic	bioclastic	crystalline
Rock	A, B, F	E	C, D

c_____

2. The diagram below shows a drill core of sediment that was taken from the bottom of a lake.

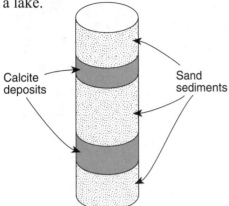

Calcite deposits

Sand sediments

Which types of rock would most likely form from compaction and cementation of these sediments?

(1) sandstone and limestone
(2) shale and coal
(3) breccia and rock salt
(4) conglomerate and siltstone   2_____

3. Which rock was organically formed and sometimes contains fossilized plant impressions?

(1) phyllite (3) rock gypsum
(2) breccia   (4) bituminous coal 3_____

4. Which rock is made up of the largest particles?

(1) conglomerate (3) shale
(2) sandstone   (4) rock salt 4_____

5. Which type of rock most likely contains fossils?

(1) scoria   (3) schist
(2) gabbro   (4) shale   5_____

6. The rounded pebbles of this rock have been cemented together to form

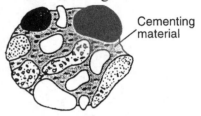

Cementing material

(Actual size)

(1) granite, an igneous rock
(2) conglomerate, a sedimentary rock
(3) siltstone, a sedimentary rock
(4) gneiss, a metamorphic rock   6_____

7. What are the rock name and map symbol used to represent the sedimentary rock that has a grain size of 0.006 to 0.2 centimeters?

Rock name: Siltstone
Map symbol:

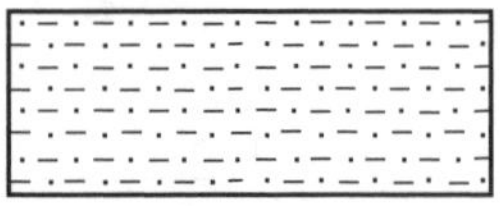

(1)

Rock name: Sandstone
Map symbol:

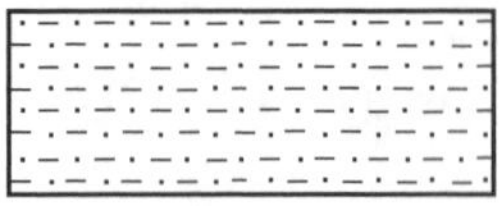

(3)

Rock name: Siltstone
Map symbol:

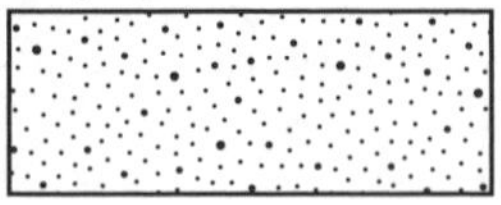

(2)

Rock name: Sandstone
Map symbol:

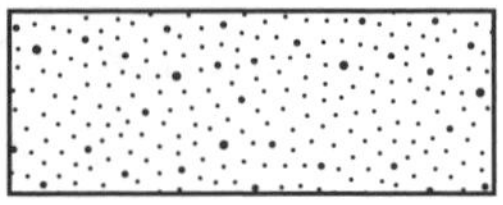

(4)

7_____

8. Give at least two processes to form a sedimentary rock.

(1)_____ (2)_____

9. Which sedimentary rock is made from the cementation and/or compaction of sediments that are less than 0.0004 cm in size? _____

10. Which sedimentary rock may form as a result of biologic processes?

    (1) shale   (3) fossil limestone
    (2) siltstone   (4) breccia   10_____

11. Dolostone is classified as which type of rock?

    (1) land-derived sedimentary rock
    (2) chemically formed sedimentary rock
    (3) nonfoliated metamorphic rock
    (4) foliated metamorphic rock   11_____

12. The block diagram below shows a portion of the Earth's crust. Letters *A*, *B*, *C*, and *D* indicate sedimentary layers.

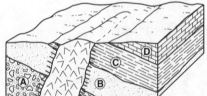
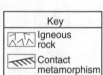

Key
Igneous rock
Contact metamorphism

    Which processes produced rock layer *B*?

    (1) subduction and melting
    (2) uplift and solidification
    (3) heat and pressure
    (4) compaction and cementation   12_____

13. Which sedimentary rocks are clastic and consist of particles that have diameters smaller than 0.005 centimeter?

    (1) conglomerate and sandstone
    (2) siltstone and shale
    (3) bituminous coal and breccia
    (4) fossil limestone and
         chemical limestone   13_____

14. Most rock gypsum is formed by the

    (1) heating of previously existing foliated bedrock
    (2) cooling and solidification of lava
    (3) compaction and cementation of shells and skeletal remains
    (4) chemical precipitation of minerals from seawater   14_____

15. Evaporite deposits could be composed of which minerals?

    (1) garnet and pyroxene
    (2) mica and feldspar
    (3) hornblende and olivine
    (4) halite and gypsum   15_____

16. The diagram below shows some features in a cave.

    Which type of rock was chemically weathered by acidic groundwater to produce the cave and its features?

    (1) siltstone   (3) quartzite
    (2) basalt   (4) limestone   16_____

17. The precipitation of the mineral halite would form a layer of

    (1) limestone   (3) coal
    (2) rock salt   (4) dolostone   17_____

18. In which set are the rock drawings labeled with their correct rock types?

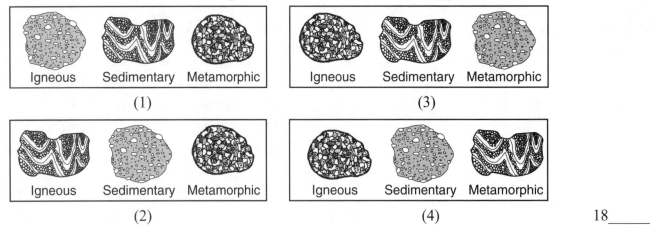

(1)

(3)

(2)

(4)

18 _____

19. The accompanying profile shows the average diameter of sediment that was sorted and deposited in specific areas *A*, *B*, *C*, and *D* by a stream entering an ocean.

As compaction and cementation of these sediments eventually occur, which area will become sandstone?

(1) *A*   (2) *B*   (3) *C*   (4) *D*

19 _____

20. The sequence of diagrams below represents the gradual geologic changes in layer **X**, located just below Earth's surface.

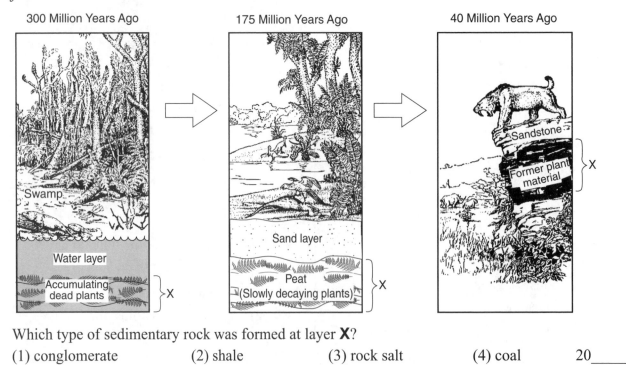

Which type of sedimentary rock was formed at layer **X**?

(1) conglomerate   (2) shale   (3) rock salt   (4) coal

20 _____

Scheme for Sedimentary Rock Identification

21. The accompanying diagram represents the fossils found in a bedrock formation located in central New York State. In which type of rock were the fossils most likely found?

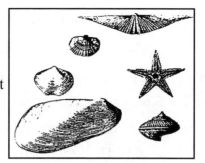

(1) sedimentary rock that formed in an ocean environment
(2) sedimentary rock that formed in a land environment
(3) igneous rock that formed in an ocean environment
(4) igneous rock that formed in a land environment

21 _____

Base your answers to question 22 on the Rock Classification Flowchart shown below. Letters *A*, *B*, and *C* represent specific rocks in this classification scheme.

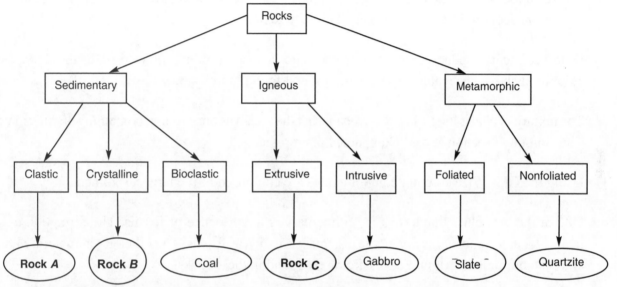

22. *a*) Rock *A* is composed of fine-grained quartz and feldspar particles 0.08 cm in diameter. State the name of Rock *A*. _____

b) Rock *B* reacts with hydrochloric acid. State the name of Rock *B*. _____

c) Which rock could also be placed where coal is positioned? _____

23. Fossils are almost always found in sedimentary rocks. Give one reasons why sedimentary rocks tend to contain fossils. _____

24. List *two* processes that would change the accumulated sediments in a delta into sedimentary rocks.

(1)_____   (2)_____

25. List *two* rocks classified as an evaporate. (1)_____ (2)_____

26. A sedimentary rock sample has the same basic mineral composition as granite. Describe *one* observable characteristic of the sedimentary rock that is different from granite.

_____

_____

Scheme for Sedimentary Rock Identification

Set 1 – Answers

1. *a*) 2 All six rocks are sedimentary in origin. Most sedimentary rocks are formed by compaction and/or cementation of sediments in a water environment. This is shown on the Rock Cycle chart.

   b) 3 Open to the Sedimentary Rock chart. The upper chart is the Inorganic Land-Derived Sedimentary Rock section. The Composition section lists the different minerals that make up these five rocks. Quartz, feldspar and clay minerals are listed here and would be found in sandstone and shale.

   c) 1 Open to the Sedimentary Rock chart and locate the three different textures. Rocks *A*, *B*, *C*, and *D* are all located in the clastic texture section. Limestone (E) is found in the crystalline or bioclastic texture section. Rock salt (F) is crystalline in texture.

2. 1 Sand sediments produce sandstone, and calcite deposits produce limestone. In the Composition column, it shows that calcite is the main mineral found in limestone.

3. 4 The texture of coal is bioclastic. This means that the rock was once living, containing organic material. The Comment section states "Compacted plant remains."

4. 1 Using the Grain Size column, conglomerates are composed of the largest sediments.

5. 4 One of the best clues that a rock is sedimentary is the presence of fossils. The formation of shale in a water environment may preserve organism remains as fossils. The other given rocks are igneous and metamorphic in origin. Fossils are rarely found in these rocks.

6. 2 The sediments in the diagram are actual size. These cemented, large sediments form conglomerates.

7. 4 In the Grain Size column, 0.006 to 0.2 cm size sediments are classified as sand. Choice 4 is the correct map symbol for sandstone.

8. Answer: Credit is award for listing any 2 of the following answers:

   deposition – burial – compaction – cementation – evaporation – precipitation

   Explanation: Open to the Rock Cycle chart. Here it states that sedimentary rocks are formed from sediments that have undergone deposition, burial, compaction and/or cementation. The Sedimentary Rock chart, Crystalline Texture – Comments section states "precipitates and evaporites."

9. Answer: Shale

   Explanation: The grain (sediment) size less than 0.0004 cm falls in the given grain size for shale.

Scheme for Metamorphic Rock Identification

TEXTURE		GRAIN SIZE	COMPOSITION	TYPE OF METAMORPHISM	COMMENTS	ROCK NAME	MAP SYMBOL
FOLIATED	MINERAL ALIGNMENT	Fine	MICA · QUARTZ · FELDSPAR · AMPHIBOLE · GARNET · PYROXENE	Regional (Heat and pressure increases)	Low-grade metamorphism of shale	Slate	
		Fine to medium			Foliation surfaces shiny from microscopic mica crystals	Phyllite	
					Platy mica crystals visible from metamorphism of clay or feldspars	Schist	
	BAND-ING	Medium to coarse			High-grade metamorphism; mineral types segregated into bands	Gneiss	
NONFOLIATED		Fine	Carbon	Regional	Metamorphism of bituminous coal	Anthracite coal	
		Fine	Various minerals	Contact (heat)	Various rocks changed by heat from nearby magma/lava	Hornfels	
		Fine to coarse	Quartz	Regional or contact	Metamorphism of quartz sandstone	Quartzite	
			Calcite and/or dolomite		Metamorphism of limestone or dolostone	Marble	
		Coarse	Various minerals		Pebbles may be distorted or stretched	Metaconglomerate	

Overview:

Metamorphic rocks have been produced from preexisting rocks by the addition of heat and/or pressure (metamorphism). In the formation of metamorphic rocks, physical and chemical changes occur but melting never occurred. Regional metamorphism is associated with a large area of metamorphism, formed by the pressure and heat of colliding plates. Contact metamorphism is when the heat of magma/lava touches rocks, causing a small area to change into metamorphic rocks. Contact metamorphism will occur on the outside edges of igneous intrusions. It is here where the heat of the molten material comes into contact with existing layers of rocks. In both types of metamorphism, the preexisting rock is changed to a new metamorphic rock, having different properties.

The Chart:

This chart is divided into two sections based on texture type. The upper section has a texture labeled Foliated, while the lower section of this chart is labeled Nonfoliated. Foliated texture is where a metamorphic rock exhibits a "layering" effect.

Upper Chart:

Foliated Texture section – Foliated texture is subdivided into Mineral Alignment and Banding. In Mineral Alignment texture, the pressure has caused the minerals to line up, thus the term mineral alignment. Banding texture, associated with the metamorphic rock gneiss, is when minerals are lined up or segregated in a "wavy band."

Composition – The shaded bars represent the different minerals found in the four metamorphic rocks shown to the right. As shown, slate's composition is only mica, while gneiss would contain all six minerals given in this composition section.

Type of Metamorphism – All of the rocks listed in this upper chart were formed by regional metamorphism. Regional Metamorphism is a large area that has undergone metamorphic change. It is associated with plate tectonics and the formation of folded mountains caused by colliding plates. The downward pointing arrow indicates that, as pressure and heat increase, the rock changes from slate to phyllite to schist, and finally schist changes to gneiss.

Comments – Use these comments to help identify an unknown metamorphic rock. Notice that slate originated from shale, a sedimentary rock. Also, banding is described in the Gneiss Comments section.

Map Symbol – Shown in this column are the designated diagrammed symbols for the given rocks.

Lower Chart:

Nonfoliated Texture section – Nonfoliated texture is when the minerals have not lined-up due to metamorphism.

Composition – Coal is made from the element carbon. Quartz is the dominant mineral found in quartzite. Marble is composed mainly of calcite and/or dolomite.

Types of Metamorphism – Both types are listed here: contact and regional metamorphism. Hornfels is associated only with contact metamorphism. The last three rocks are associated with both types of metamorphism.

Comments – In the comment section for hornfels, a simple definition of contact metamorphism is given: "Various rocks changed by heat from nearby magma/lava." Sandstone, having much quartz, changes to quartzite. Limestone or dolostone changes to marble during metamorphism.

Additional Information:

- Calcite is very reactive with an acid. Calcite is the main mineral in limestone. When limestone undergoes metamorphism, it changes to marble. Thus, both limestone and marble will react to acids because they both contain the mineral calcite.

- If too much heat is applied to a rock and it melts, eventually, upon cooling, the resulting rock will always be igneous.

- Metamorphic rocks are commonly found in mountainous regions.

- As a fuel, the metamorphic anthracite coal is superior to the sedimentary bituminous coal. Anthracite coal produces more heat and burns cleaner than other types of coal.

- Recrystallization is the growth of new crystals from the minerals already present. It is caused by the heat/pressure of metamorphism.

Diagrams:

1. **Gneiss** – Gneiss is a metamorphic rock that formed from a preexisting rock that had been subjected to great pressure and heat due to regional metamorphism. These conditions caused the minerals to segregate into "wavy" bands. This is very evident in the sketch of gneiss (1). The actual photograph (2) of this banded metamorphic rock clearly shows the segregation of the minerals.

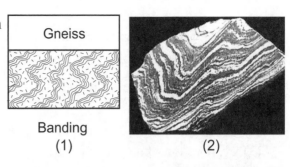

Gneiss

Banding
(1)

(2)

2. **Foliated Texture** – Diagram 1 represents the igneous rock granite, having interlocking crystals. When it undergoes metamorphism, the increase of heat and pressure cause minerals to undergo alignment producing a foliated texture as shown in diagram 2. Slate, phyllite, and schist are rocks that are classified in this texture section of the Metamorphic Rock Identification chart.

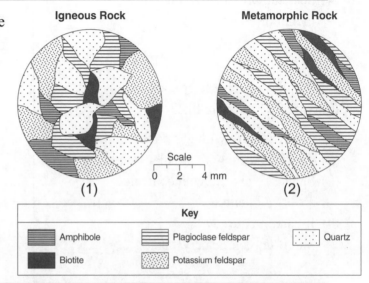

Igneous Rock

Metamorphic Rock

Scale
0 2 4 mm

(1)

(2)

Key		
▤ Amphibole	▤ Plagioclase feldspar	⋯ Quartz
■ Biotite	▨ Potassium feldspar	

3. **Contact Metamorphism** – Letter *B* represents an intrusion of magma that solidified into an igneous rock. When the magma intruded the rock layers, the heat of the magma caused contact metamorphism to the previous existing rocks. In the contact area, letter *D*, sandstone, is now quartzite; letter *C*, shale, is now slate; and letter *A*, limestone, has been metamorphosed to marble.

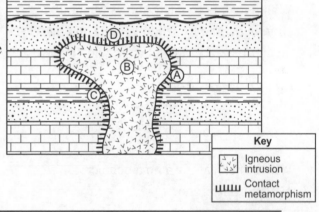

Key	
⋁⋀	Igneous intrusion
⊔⊔⊔⊔	Contact metamorphism

4. **Sedimentary to Metamorphic** – Bituminous coal is a sedimentary rock formed by compaction of organic matter. If this rock layer is subjected to folding, due to plate tectonics causing regional metamorphism, an increase of heat and pressure changes the soft bituminous coal into hard anthracite coal. This metamorphic coal is the highest grade form of coal.

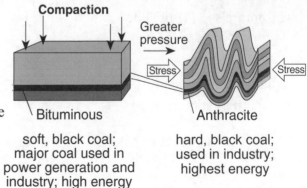

Compaction

Greater pressure

Stress   Stress

Bituminous

soft, black coal; major coal used in power generation and industry; high energy

Anthracite

hard, black coal; used in industry; highest energy

Set 1 — Scheme for Metamorphic Rock Identification

1. Wavy bands of light and dark minerals visible in gneiss bedrock probably formed from the

   (1) cementing together of individual mineral grains
   (2) cooling and crystallization of magma
   (3) evaporation of an ancient ocean
   (4) heat and pressure during metamorphism 1 _______

2. Which physical characteristic best describes the rock phyllite?

   (1) glassy texture with gas pockets
   (2) clastic texture with angular fragments
   (3) bioclastic texture with cemented shell fragments
   (4) foliated texture with microscopic mica crystals 2 _______

3. Which rock is foliated, shows mineral alignment but not banding, and contains medium-sized grains of quartz and pyroxene?

   (1) phyllite (3) gneiss
   (2) schist (4) quartzite 3 _______

4. Which rock can form in a contact metamorphic zone?

   (1) slate (3) gneiss
   (2) hornfels (4) phyllite 4 _______

5. Which metamorphic rock would contain more different minerals?

   (1) gneiss (3) marble
   (2) slate (4) phyllite 5 _______

6. How do the metamorphic rocks schist and quartzite differ?

   (1) Quartzite contains the mineral quartz and schist does not.
   (2) Quartzite forms from regional metamorphism and schist does not.
   (3) Schist is organically formed and quartzite is not.
   (4) Schist is foliated and quartzite is not 6 _______

7. Which rock would most likely be produced by the metamorphism of gray limestone?

   (1) quartzite (3) marble
   (2) slate (4) gneiss 7 _______

8. The diagram below shows a portion of the Earth's crust. Letters *A*, *B*, *C*, and *D* indicate different types of rock.

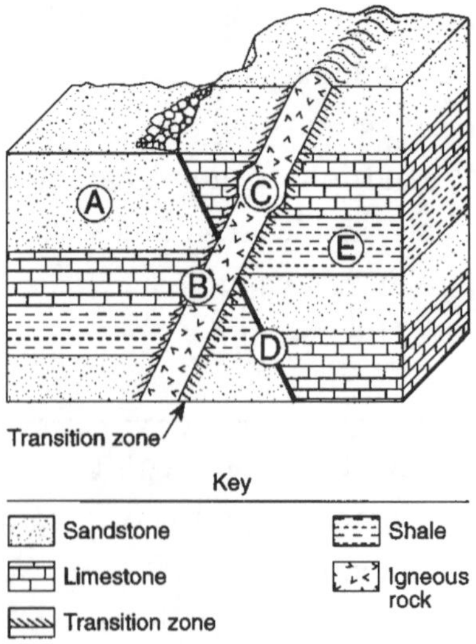
A
B
C
D
E
Transition zone
Key
Sandstone
Shale
Limestone
Igneous rock
Transition zone

At which location is metamorphic rock most likely to be found?

   (1) *A* (3) *C*
   (2) *B* (4) *D* 8 _______

9. Which rock forms by the recrystallization of unmelted rock material under conditions of high temperature and pressure?

    (1) granite (3) rock gypsum
    (2) gneiss   (4) bituminous coal   9 _____

10. The crystals of many metamorphic rocks are aligned in bands as a result of

    (1) earthquake faulting
    (2) cooling and solidification
    (3) mechanical weathering
    (4) heat and pressure   10_____

11. The diagram below shows the geologic cross section. Location *A* is within the metamorphic rock.

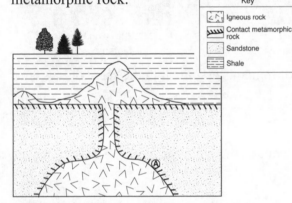

Key
Igneous rock
Contact metamorphic rock
Sandstone
Shale

The metamorphic rock at location *A* is most likely

    (1) marble   (3) phyllite
    (2) quartzite   (4) slate   11 _____

12. Which metamorphic rock experienced the lowest-grade of metamorphism?

    (1) slate   (3) gneiss
    (2) schist   (4) quartzite   12 _____

13. The heat and pressure of regional metamorphism might affect a conglomerate layer by distorting or stretching the pebbles. The resulting rock would be

    (1) hornfels   (3) phyllite
    (2) granite   (4) metaconglomerate
                                    13_____

14. Slate is formed by the

    (1) deposition of chlorite and mica
    (2) foliation of schist
    (3) metamorphism of shale
    (4) folding and faulting of gneiss 14 _____

15. What metamorphic rock would react with hydrochloric acid?

    (1) slate   (3) marble
    (2) quartzite   (4) phyllite   15 _____

16. Which rock was subjected to intense heat and pressure but did not solidify from magma?

    (1) sandstone   (3) gabbro
    (2) phyllite   (4) rhyolite   16_____

17. The photograph shows a large outcrop of rock composed primarily of visible crystals of mica, quartz, and feldspar.

Based on the composition and foliated texture, this rock can best be identified as

    (1) marble   (3) slate
    (2) schist   (4) anthracite coal   17 _____

18. Which sequence of change in rock type occurs as shale is subjected to increasing heat and pressure?

    (1) shale → schist → phyllite → slate → gneiss
    (2) shale → slate → phyllite → schist → gneiss
    (3) shale → gneiss → phyllite → slate → schist
    (4) shale → gneiss → phyllite → schist → slate   18_____

19. Which rock sample is most likely metamorphic rock?

        (1)   (2)   (3)   (4)   19_____

        Base your answers to question 20 on the diagram below which represent two magnified
    views showing the arrangement of minerals before and after metamorphism of a rock.

20. *a)* State the name of Rock *C*.

    _____

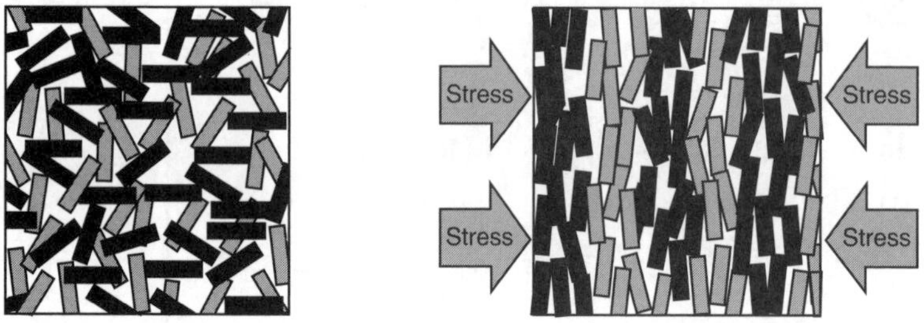

                        Mineral Arrangement   **Rock C** Showing Banding
                        Before Metamorohism   After Metamorohism

    b) Name two processes needed to cause the banding of minerals as shown in the second diagram.
        (1)_____   (2) _____

21. An igneous intrusion entered a layer of dolostone. The contact metamorphism of the dolostone would
    produce what metamorphic rock?
                                    _____

22. State the name of the rock, formed by
    contact metamorphism;

    at location *A*_____

    at location *B*_____.

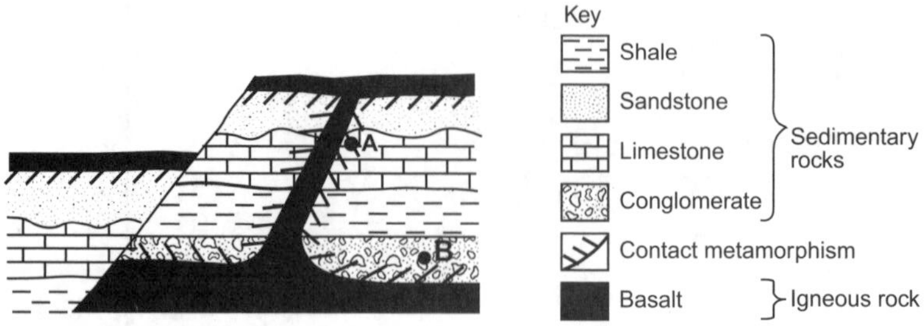

Key
Shale
Sandstone
Limestone   } Sedimentary rocks
Conglomerate
Contact metamorphism
Basalt   } Igneous rock

23. When schist is exposed to greater heat and pressure during
    metamorphism, it would change to what other metamorphic rock? _____

24. Identify *one* similarity and *one* difference between the metamorphic rocks slate and phyllite.

    Similarity _____   Difference _____

Scheme For Metamorphic Rock Identification

Answers

Set 1

1. 4 Open to the Metamorphic Rock chart and locate gneiss. In the Texture section it states Banding. In banding, minerals are segregated into wavy bands of light and dark minerals. Being a metamorphic rock, it must have been subjected to an increase in heat and pressure during metamorphism. Choices 1 and 3 are sedimentary processes, while choice 2 is an igneous process.

2. 4 In the Metamorphic Rock chart, locate phyllite. The given texture for this rock is foliated. The Comments section states that phyllite has microscopic mica crystals.

3. 2 In the Composition column, as shown by the dark bars, only schist and gneiss contain quartz and pyroxene. The Texture column gives the texture of schist being foliated with mineral alignment. Banding texture is seen in gneiss.

4. 2 In the Metamorphic Rock chart, of the given choices, only hornfels has contact as a type of metamorphism.

5. 1 In the Composition column, gneiss contains all six minerals as shown by the darken bars, while phyllite contains five minerals. The other two given rocks only contain one or two minerals.

6. 4 Open to the Metamorphic Rock chart and locate quartzite and schist. Starting with choice 1 and using this chart, see if the given statements are true or not. Using this procedure of elimination, only choice 4 is correct as shown by the Texture column.

7. 3 Limestone changes into marble when it undergoes metamorphism. This information is given in the Metamorphic Rock chart within the Comments section next to marble.

8. 2 Contact metamorphism occurs by the heat of magma/lava. From this diagram, contact metamorphism would occur along the outer edges of the igneous intrusion as it comes into contact with the rock strata. At position B, the adjacent limestone would be metamorphized into marble.

Geologic History of New York State

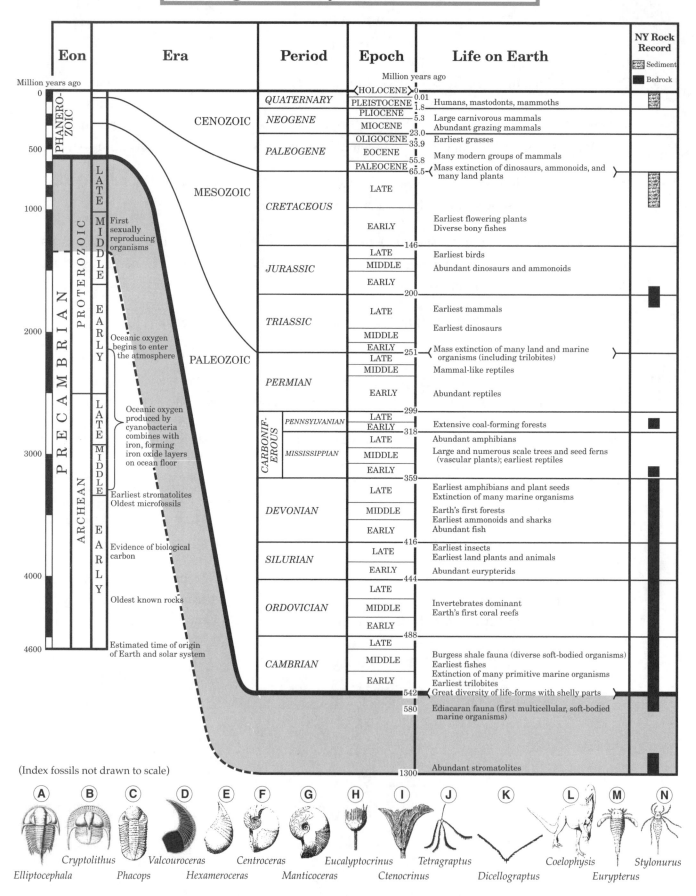

Eon	Era	Period	Epoch	Life on Earth	NY Rock Record

NY Rock Record: ▦ Sediment ■ Bedrock

Million years ago

Million years ago

PHANEROZOIC	CENOZOIC	QUATERNARY	HOLOCENE — 0		Sediment
			PLEISTOCENE — 0.01 / 1.8	Humans, mastodonts, mammoths	
		NEOGENE	PLIOCENE — 5.3	Large carnivorous mammals	
			MIOCENE — 23.0	Abundant grazing mammals	
		PALEOGENE	OLIGOCENE — 33.9	Earliest grasses	Sediment
			EOCENE — 55.8	Many modern groups of mammals	
			PALEOCENE — 65.5	Mass extinction of dinosaurs, ammonoids, and many land plants	
	MESOZOIC	CRETACEOUS	LATE		
			EARLY	Earliest flowering plants / Diverse bony fishes	
		JURASSIC — 146	LATE	Earliest birds	
			MIDDLE	Abundant dinosaurs and ammonoids	
			EARLY — 200		
		TRIASSIC	LATE	Earliest mammals / Earliest dinosaurs	
			MIDDLE		
			EARLY — 251	Mass extinction of many land and marine organisms (including trilobites)	
	PALEOZOIC	PERMIAN	LATE		
			MIDDLE	Mammal-like reptiles	
			EARLY — 299	Abundant reptiles	
		CARBONIFEROUS — PENNSYLVANIAN	LATE	Extensive coal-forming forests	
			EARLY — 318		
		MISSISSIPPIAN	LATE	Abundant amphibians	
			MIDDLE	Large and numerous scale trees and seed ferns (vascular plants); earliest reptiles	
			EARLY — 359		
		DEVONIAN	LATE	Earliest amphibians and plant seeds / Extinction of many marine organisms	
			MIDDLE	Earth's first forests / Earliest ammonoids and sharks	
			EARLY — 416	Abundant fish	
		SILURIAN	LATE	Earliest insects / Earliest land plants and animals	
			EARLY — 444	Abundant eurypterids	
		ORDOVICIAN	LATE		
			MIDDLE	Invertebrates dominant / Earth's first coral reefs	
			EARLY — 488		
		CAMBRIAN	LATE	Burgess shale fauna (diverse soft-bodied organisms)	
			MIDDLE	Earliest fishes / Extinction of many primitive marine organisms	
			EARLY — 542	Earliest trilobites / Great diversity of life-forms with shelly parts	

Precambrian (left side):

PRECAMBRIAN — PROTEROZOIC — ARCHEAN

LATE / MIDDLE / EARLY (Proterozoic)
LATE / MIDDLE / EARLY (Archean)

- First sexually reproducing organisms
- Oceanic oxygen begins to enter the atmosphere
- Oceanic oxygen produced by cyanobacteria combines with iron, forming iron oxide layers on ocean floor
- Earliest stromatolites / Oldest microfossils
- Evidence of biological carbon
- Oldest known rocks
- Estimated time of origin of Earth and solar system

— 580 Ediacaran fauna (first multicellular, soft-bodied marine organisms)

— 1300 Abundant stromatolites

(Index fossils not drawn to scale)

A	B	C	D	E	F	G	H	I	J	K	L	M	N
Elliptocephala	Cryptolithus	Phacops	Valcouroceras / Hexameroceras	Centroceras		Eucalyptocrinus / Manticoceras	Ctenocrinus	Tetragraptus		Dicellograptus	Coelophysis	Eurypterus	Stylonurus

Geologic History of New York State

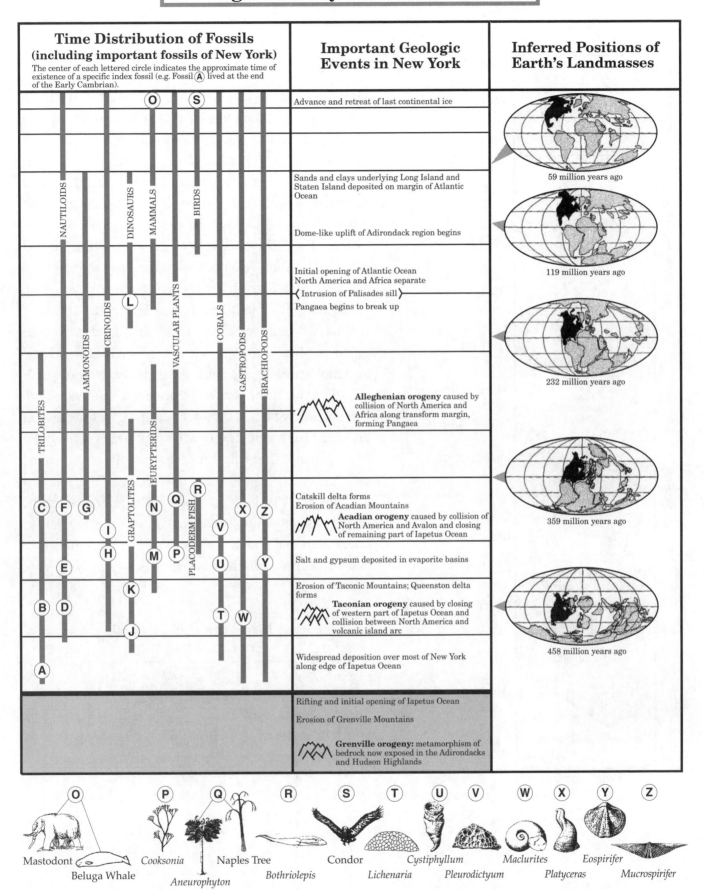

Time Distribution of Fossils
(including important fossils of New York)

The center of each lettered circle indicates the approximate time of existence of a specific index fossil (e.g. Fossil Ⓐ lived at the end of the Early Cambrian).

Fossils on chart: NAUTILOIDS, AMMONOIDS, CRINOIDS, DINOSAURS, MAMMALS, VASCULAR PLANTS, BIRDS, CORALS, GASTROPODS, BRACHIOPODS, TRILOBITES, EURYPTERIDS, GRAPTOLITES, PLACODERM FISH

Index fossils: Ⓒ Ⓕ Ⓖ Ⓘ Ⓗ Ⓔ Ⓚ Ⓑ Ⓓ Ⓙ Ⓐ Ⓛ Ⓝ Ⓠ Ⓡ Ⓜ Ⓟ Ⓥ Ⓤ Ⓣ Ⓧ Ⓩ Ⓨ Ⓦ Ⓞ Ⓢ

Important Geologic Events in New York

Advance and retreat of last continental ice

Sands and clays underlying Long Island and Staten Island deposited on margin of Atlantic Ocean

Dome-like uplift of Adirondack region begins

Initial opening of Atlantic Ocean
North America and Africa separate
⟨ Intrusion of Palisades sill ⟩
Pangaea begins to break up

Alleghenian orogeny caused by collision of North America and Africa along transform margin, forming Pangaea

Catskill delta forms
Erosion of Acadian Mountains
Acadian orogeny caused by collision of North America and Avalon and closing of remaining part of Iapetus Ocean

Salt and gypsum deposited in evaporite basins

Erosion of Taconic Mountains; Queenston delta forms
Taconian orogeny caused by closing of western part of Iapetus Ocean and collision between North America and volcanic island arc

Widespread deposition over most of New York along edge of Iapetus Ocean

Rifting and initial opening of Iapetus Ocean

Erosion of Grenville Mountains

Grenville orogeny: metamorphism of bedrock now exposed in the Adirondacks and Hudson Highlands

Inferred Positions of Earth's Landmasses

59 million years ago

119 million years ago

232 million years ago

359 million years ago

458 million years ago

Ⓞ Mastodont, Beluga Whale
Ⓟ Cooksonia
Ⓠ Naples Tree, Aneurophyton
Ⓡ Bothriolepis
Ⓢ Condor
Ⓣ Cystiphyllum, Lichenaria
Ⓤ Pleurodictyum
Ⓥ Maclurites
Ⓦ Platyceras
Ⓧ Eospirifer
Ⓨ Mucrospirifer
Ⓩ

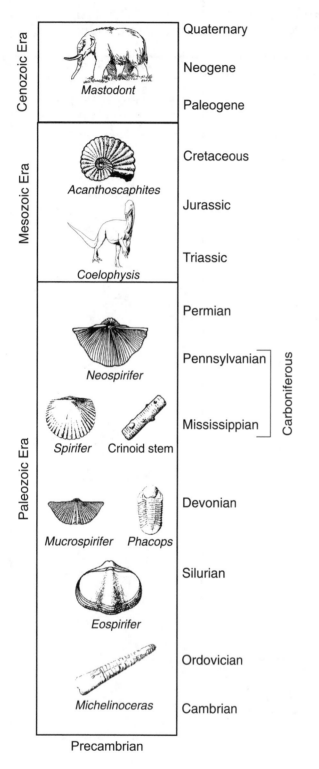

Cenozoic Era

Quaternary

Neogene

Paleogene

Mastodont

Mesozoic Era

Cretaceous

Acanthoscaphites

Jurassic

Triassic

Coelophysis

Paleozoic Era

Permian

Pennsylvanian ⎤
               ⎥ Carboniferous
Mississippian ⎦

Neospirifer

Spirifer   *Crinoid stem*

Devonian

Mucrospirifer   *Phacops*

Silurian

Eospirifer

Ordovician

Cambrian

Michelinoceras

Precambrian

Overview:

Using radioactive dating, geologists tell us that the Earth is close to 4.6 billion years in age. By studying and dating fossils, a time line has been created detailing the great diversity of life forms this planet has supported. Fossils have also revealed that periods of major extinction have occurred throughout the history of our planet. Yet, during most of this 4.6 billion year time span, very little life existed. The earliest and longest division of time, lasting close to 4 billion years is named the Precambrian Eon. Very little fossilized life has been found from this eon. Then, around 542 millions of years ago (mya), abundant multi-cell sea life appeared. This marked the start of the Paleozoic Era – "the age of sea-life." As fossil evidence accumulated, detailing episodes of mass extinction along with the appearance of new life, eras were subdivided into smaller units of time called periods and epochs. The next era is the Mesozoic Era, in which dinosaurs were "king of the planet." The present era, the one we live in, is named the Cenozoic Era – "the age of mammals." As one can see, this chart is the largest one in the reference tables, containing much information. We will not attempt to cover it all, but enough of it so you can be comfortable and successful with the Geologic History of New York State chart.

**New York State Fossil
Eurypterus remipes**

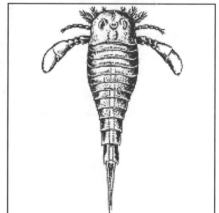

The Chart:

Precambrian Eon - Archean and Proterozoic – On the far left is the main time line. At the bottom it states, "Estimated time of origin of Earth and solar system" and is dated 4,600 million years ago, which equals 4.6 billion years. This starts the Precambrian Eon, which is divided into two sections – the older Archean and the younger Proterozoic Eon. Next to each is important information on major events that occurred during this large time span. Notice the gray area, representing the upper Proterozoic, extends across the bottom of the chart and contains additional information. Because not much is known about the Precambrian, relatively little information is given for an eon that lasted close to 4 billion years, covering almost 90% of the entire geologic history of our planet. Yet, during the Middle Archean, oxygen was being produced by cyanobacteria, a blue-green algae. Eventually, as oxygen increased in the ocean, some of it escaped into the atmosphere. The Sun's rays reacted with the oxygen changing some of the oxygen (O_2) to ozone (O_3), which started filtering out harmful UV rays. Conditions in the ocean, and eventually in the atmosphere, became favorable for life, and sea life exploded bringing an end to the Precambrian Eon and introducing the Paleozoic Era.

Paleozoic Era

Eras - Periods - Epochs – These divisions of the geologic time are based on fossil evidence. On the right side of the epoch column is a time line that dates the beginning and ending of all divisions. The Paleozoic Era is divided into six periods and many smaller epochs. This era started 542 million years ago (mya) and ending 251 mya, making it the second longest division of geologic time. The ending of this era started the Mesozoic Era, which lasted 185.5 million years. In this era, dinosaurs lived along with smaller mammals. This era consists of three periods, of which Jurassic is best known to all moviegoers who watched Jurassic Park. Dinosaurs lived during all three periods, with the largest ones living during the Cretaceous Period. The present era is the Cenozoic Era, which continues today and includes three periods and seven epochs having specific names. Being the youngest and shortest era, much information has been well preserved, especially during the Pleistocene and Holocene Epochs.

Life on Earth – In this section, the different life forms are placed in the period according to when they first appeared, based on fossil evidence. In the Precambrian oceans, simple single-cellular life and later multicellular, soft-bodied marine organisms existed. The Cambrian Period, the earliest period of the Paleozoic, is when great diversity of sea life started. Moving upward through the Paleozoic, many "Earliest" species are listed. A notable one is the marine organism, the trilobite – a hard-shelled, segmented marine organism that existed throughout the Paleozoic Era. At the end of the Paleozoic Era, which is the top of the Late Permian Period, is the statement: "Mass extinction of many land and marine organisms (including trilobites)." Mass extinction closes out most geologic divisions. Eventually, new life forms are introduced. This was true for the Mesozoic Era, with the introduction of dinosaurs. Earliest mammals, birds, and earliest flowering plants also had their beginning in the Mesozoic Era. The extinction of dinosaurs, along with many other species, ended this era. Life on Earth in the Cenozoic Era is about the rise and diversification of mammals as the dominant life form. At the very top of the chart, a relatively new species appeared – Humans. This occurred in the Pleistocene Epoch. We have been here, truly, a very short time when viewed by geologic time.

Geologic History of New York State　　　　**Page 109**

New York Rock Record – A dark bar means that this period's surface bedrock is present somewhere in NYS. A light color bar indicates that sediments from a specific geologic age are found somewhere within NYS. The absence of a bar indicates that this age's bedrock/sediment is not found in NYS, indicating a time of uplift and erosion of this bedrock. For example, the Permian Period has no dark bar, meaning that this period's bedrock is not found in NYS; it has been eroded away. Notice that the rock record is well preserved for the Paleozoic Era. NYS is famous for its well-preserved Paleozoic fossils. As shown by the bars, Mesozoic rocks are few in NYS. This is the reason why no dinosaur bones have ever been found in our state, although some footprints have been preserved.

Time Distribution of Fossils – This section contains gray bars spanning time periods when certain species existed. For example, the first bar, the Trilobites bar, shows that the trilobites came into existence during the Early Cambrian Period, existing until the end of the Permian Period. The letters located on the bars indicate the approximate time when specific index fossils lived. These letters correspond to the diagrammed fossils at the bottom of this chart. Check out letter *L*. The bar shows that this dinosaur, Coelophysis, lived during the Late Triassic Period, and its footprint has been found in NYS. Notice the abundance of NYS index fossils that lived during the Paleozoic Era. One of these index fossils, letter *M*, eurypterus, is a member of the eurypterids and is our state fossil.

Important Geologic Events in New York – This section has information regarding the geologic events that occurred in NYS. The youngest geologic event found at the top of the chart is "Advance and retreat of last continental ice." This ice age event lines up with the Pleistocene Epoch. Humans were around during this event. The oldest event, located in the gray Precambrian section, is the Grenville orogeny. An orogeny is a mountain building episode. As shown by the diagrammed mountains, there were four orogenies affecting NYS. At the end of the Triassic Period the supercontinent Pangaea began to break-up. Volcanic activity related to this event produced the intrusion of the Palisades sill. This igneous intrusion is now exposed forming the cliffs along the west side of the Hudson River, across from New York City.

Inferred Positions of Earth's Landmasses – This column shows the inferred positions of the continents as they drifted due to plate tectonics. These diagrammed globes are positioned close to their respective time period, with the small gray arrow pointing to its position within the correct geologic period. Under the globes is given the time of occurrence in mya. Notice that North America was much closer to the equator 359 mya. This is evident because fossilized Devonian coral can be found throughout NYS. Remember, coral lives in warm tropical waters. With the break-up of the supercontinet Pangaea, North America has drifted northwest to its present position.

Additional Information:

- The mass extinction at the end of the Mesozoic Era was probably caused by an impact of an asteroid. This collision would have released tremendous amounts of ash and dust into the atmosphere, causing a rapid cooling of the atmosphere. The resulting colder climate would then put much stress on many species that live in warm temperature areas.

- Mastodonts are related to the mammoths. Fossilized mastodonts have been found in NYS.

- The Adirondacks are listed twice in the Geologic Events section, once in the Precambrian Eon and again in the Cretaceous Period.

- This chart is often used with the Bedrock Geology and the Landscape Regions of NYS charts.
- During the Proterozoic Eon, oxygen produced by the process of photosynthesis by cyanobacteria – one of the earliest blue-green algae to exist on our planet – slowly entered the atmosphere. Stromatolites are rock-like structures formed from the remains of cyanobacteria and the precipitation of minerals. Some fossilized stromatolites have been dated back to the Archean Eon. These fossilized remains are exposed in Saratoga County, NY.

Diagrams:

1. **Time Relationships** – According to the Big Bang theory, astronomers estimate that our universe started 13.7 billion years ago (bya). The estimated origin of our Earth and solar system is 4.6 bya. This starts the Precambrian Eon where first life – single cell organisms – appeared in the fossil record 3.6 bya. When the atmosphere became oxygen rich, multicellular, soft body organisms had a better chance of surviving in the seas and oceans. The darker upper part of this diagram represents the Paleozoic, the Mesozoic and the Cenozoic Eras. Humans first appeared, as shown, on the very top of the diagram.

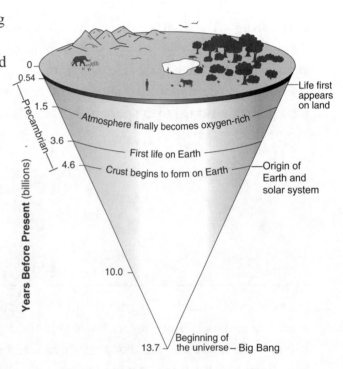

2. **Eon and Era Relationship** – This pie graph represents the percentage of total time for the four major divisions of geologic time. In time relationship, the Precambrian Eon is close to 90% of all Earth's history, followed by the Paleozoic and Mesozoic Eras. The present one, the Cenozoic Era, spans 65.5 million years, yet it is by far the shortest era. Each time division was based on fossil evidence of a major mass extinction event and/or the appearance of new life.

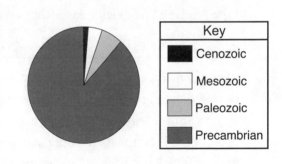

3. **Human Existence** – This pie graph represents the existence of humans compared to the 65.5 million years that the Cenozoic Era spans. In relationship to all of Earth's 4.6 billion year history, humans have lived a milli-fraction of time on this planet. In such a short time, we have had a major negative impact on the lives of other organisms, and we have overused our natural resources.

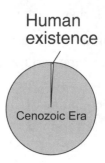

Human existence

Cenozoic Era

4. **Outgassing** – The Earth's earliest atmosphere was formed during the Archean Eon. This poisonous, non-oxygen atmosphere evolved as gases escaped from the Earth's interior – a process known as outgassing. These gases were probably similar to the poisonous gases that presently escape from volcanoes.

Outgassing

Water vapor
Carbon Dioxide
Ammonia
Nitrogen
Methane

5. **Index Fossils** – Index fossils are fossils that identify specific geologic bedrock layers. Two conditions defining index fossils are: they should be widespread, and should have lived a relatively short period of time (thus exist in few bedrock layers). According to the Time Distribution of Fossil chart and the diagrammed index fossils, layer *A* is Devonian-age, layer *B* is Silurian, and layer *C* is Cambrian. The wavy line shows an unconformity due to the missing Ordovician layer that was eroded away.

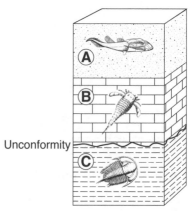

Unconformity

A
B
C

6. **Unconformity** – In an unconformity, layer(s) are missing due to erosional forces. The usual steps in the formation of an unconformity are: underwater deposition of sediments forming sedimentary layers, uplift of layers out of the water environment, erosion of specific layers, subsidence (sinking or lowering of the land until under water), and deposition of new layers. In this unconformity, the Silurian layer is missing being eroded away. The igneous intrusion is the youngest event shown as it cuts through all layers, with contact metamorphism occurring to all layers.

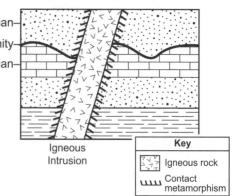

Devonian
Unconformity
Ordovician

Igneous Intrusion

Key

Igneous rock

Contact metamorphism

Geologic History of New York State

1. Which column best represents the relative lengths of time of the major intervals of geologic history?

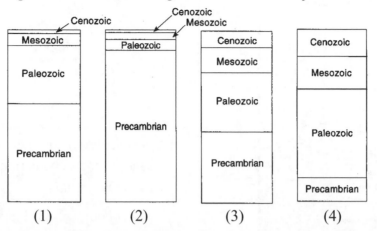

(1)   (2)   (3)   (4)   1 _____

2. The geologic drill core below shows bedrock layers A, B, and C that have not been overturned. The geological age of layer B is

(1) Cambrian
(2) Devonian
(3) Ordovician
(4) Permian   2 _____

Top
A Carboniferous
B
C Silurian   Bottom

3. The index fossil below was found in surface bedrock in New York State. This index fossil is representative of a group of invertebrate animals known as

(1) trilobites
(2) stromatolites
(3) brachiopods
(4) eurypterids   3 _____

4. Which important New York State fossil is most likely to be found in the Triassic Age rocks in the Newark Lowlands?

(1) eurypterid
(2) mastodont
(3) coelophysis
(4) naples tree   4 _____

5. Trilobite fossils were recently discovered in Himalayan Mountain bedrock. During which geologic period could this bedrock have been formed?

(1) Neogene   (3) Triassic
(2) Cretaceous   (4) Cambrian   5 _____

6. During which geologic time period was Stylonurus alive and abundant?

(1) Cambrian   (3) Silurian
(2) Jurassic   (4) Devonian   6 _____

7. Which major mountain-building episode is most recent?

(1) Grenville orogeny
(2) Taconian orogeny
(3) Acadian orogeny
(4) Alleghenian orogeny   7 _____

8. During which era did the initial opening of the present-day Atlantic Ocean most likely occur?

(1) Cenozoic   (3) Late Proterozoic
(2) Mesozoic   (4) Paleozoic   8 _____

The cartoon below illustrates possible interaction between humans and mammoths.

The primitive game of "Pull the mammoth's tail and run"

9. During which geologic timespan could this "game" have occurred?

   (1) Pleistocene Epoch
   (2) Pennsylvanian Epoch
   (3) Precambrian Eon
   (4) Paleozoic Era   9 _____

10. During which geologic epoch did the glacier retreat from New York State?

   (1) Pleistocene
   (2) Eocene
   (3) Late Pennsylvanian
   (4) Early Mississippian   10 _____

11. According to the Geologic History of New York State in the Earth Science Reference Tables, the inferred latitude of New York State 359 million years ago was closest to

   (1) where it is now
   (2) the North Pole
   (3) the Equator
   (4) 45° south   11 _____

Base your answers to question 12 on the chart below, which shows the geologic ages of some well known fossils

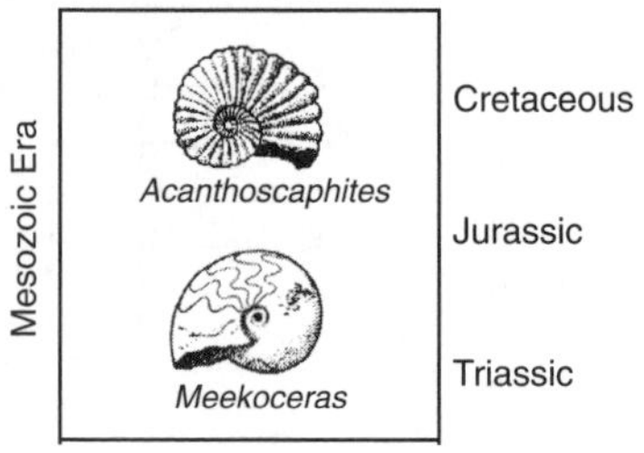

Mesozoic Era

Acanthoscaphites — Cretaceous

— Jurassic

Meekoceras — Triassic

12. Which New York State fossil is found in rocks of the same period of geologic history as Meekoceras?

   (1) Condor   (3) Placoderm fish
   (2) Eurypterus   (4) Coelophysis 12 _____

13. Bedrock of which four consecutive geologic periods is best preserved in New York State?

   (1) Cambrian, Ordovician, Silurian, Devonian
   (2) Devonian, Carboniferous, Permian, Triassic
   (3) Permian, Triassic, Jurassic, Cretaceous
   (4) Jurassic, Cretaceous, Paleogene, Quaternary   13 _____

14. Which fossil would most likely be found in the same siltstone layer as a Cryptolithus fossil?

   (1)   (2)   (3)   (4)

                                       14 _____

15. The New York State index fossil Valcouroceras is classified as a

   (1) coral   (3) eurypterid
   (2) crinoid   (4) nautiloid   15 _____

16. The time line below represents the entire geologic history of Earth.

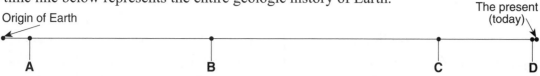

Origin of Earth

The present (today)

A   B   C   D

Which letter best represents the first appearance of humans on Earth?
(1) A   (2) B   (3) C   (4) D   16_____

17. Which sequence of New York State index fossils shows the order in which the organisms appeared on Earth?

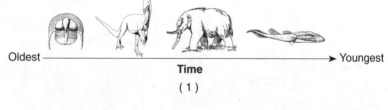

Oldest ──────────────────────→ Youngest
Time
(1)

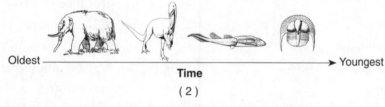

Oldest ──────────────────────→ Youngest
Time
(2)

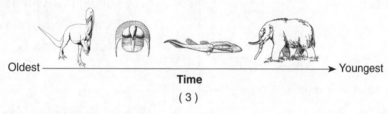

Oldest ──────────────────────→ Youngest
Time
(3)

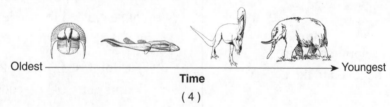

Oldest ──────────────────────→ Youngest
Time
(4)   17_____

18. State one tectonic event affecting North America that occurred near the same time as the intrusion of the Palisade Sill. _____

19. According to the geologic record, during which geologic time period were the sands and clays underlying Long Island and Staten Island deposited? _____

20. Identify by name the oldest New York State index fossil that could be found in the Early Ordovician bedrock. _____

21. What event is generally accepted as the cause of mass extinction that occurred 65.5 million years ago? _____

22. Trilobite fossil remains are most likely to be found in bedrock of

    (1) Precambrian age near Mt. Marcy
    (2) Cretaceous age on Long Island
    (3) Cretaceous age northwest of New York City
    (4) Ordovician age near Plattsburgh   22 _____

23. Which mountain range resulted from the collision of North America and Africa, as parts of Pangaea joined together in the Permian Period?

    (1) Alleghenian Mountains
    (2) Acadian Mountains
    (3) Taconic Mountains
    (4) Grenville Mountains   23 _____

24. The primitive lobe-finned fish shown below is thought to be an ancestor of early amphibians.

    This evolutionary development from fish to amphibian is believed to have occurred during the

    (1) Triassic Period
    (2) Devonian Period
    (3) Tertiary Period
    (4) Permian Period   24 _____

25. Which group of organisms, some of which were preserved as fossils in early Paleozoic rocks, are still in existence today?

    (1) brachiopods   (3) graptolites
    (2) eurypterids   (4) trilobites   25 _____

26. Based on the theory of plate tectonics, it is inferred that over the past 250 million years North America has moved toward the

    (1) northwest   (3) southeast
    (2) southwest   (4) northeast   26 _____

27. According to available fossil evidence, which set of events is listed in the correct order from earliest to most recent?

    (1) extinction of trilobites, appearance of earliest fishes, extinction of dinosaurs
    (2) appearance of first corals, appearance of earliest insects, abundant reptiles
    (3) appearance of dinosaurs, appearance of earliest amphibians, appearance of earliest grasses
    (4) abundant eurypterids, appearance of earliest birds, appearance of first coral reefs   27 _____

28. Which group of organisms is inferred to have existed for the least amount of time in geologic history?

    (1) trilobites (3) eurypterids
    (2) dinosaurs (4) placoderm fish 28 _____

29. A very large, circular, impact crater under the coast of Mexico is believed to be approximately 65.5 million years old. This impact event is inferred to be related to the

    (1) appearance of the earliest trilobites
    (2) advance and retreat of the last continental ice sheet
    (3) extinction of the dinosaurs
    (4) formation of Pangaea   29 _____

Base your answers to question 30 on the diagram which shows a portion of a geologic time line. Letters *A* through *D* represent the time intervals between the labeled events, as estimated by some scientists.

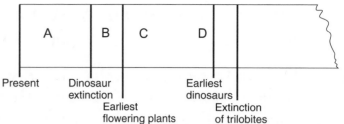

30. *a)* Fossil evidence indicates that the earliest birds developed during which time interval?

(1) *A* (2) *B* (3) *C* (4) *D* a_____

b) Earliest mammals developed during which time interval?

(1) *A* (2) *B* (3) *C* (4) *D* b_____

c) Abundant grazing mammals flourished during which time interval?

(1) *A* (2) *B* (3) *C* (4) *D* c_____

Base your answers to question 31 on the geologic time line shown below.

Geologic Time Line (millions of years ago)

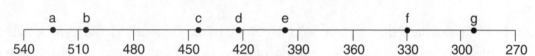

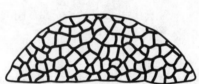

31. *a)* Place an **X** on the geologic time line, so that the center of the **X** shows the time that the coral index fossil *Lichenaria* shown above existed on Earth.

b) Letter *a* indicates a specific time during which geologic period? _____Period

c) Identify the mountain building event (orogeny) that was occurring in eastern North America at the time represented by letter *g*. _____

d) Identify *one* letter that indicates a time for which there is no rock record in New York State. _____

e) Name a brachiopod that existed during the time indicated by letter *d*. _____

32. Identify *one* important geologic event that occurred in New York State when mastodonts existed.

_____

33. Identify the eon during which the oldest microfossils occurred. _____

34. In what period did the NYS salt layers develop? _____

1. 2   The relative lengths of the different intervals of geologic history are shown on the left side of the Geologic History of NYS chart. The Precambrian Eon, by far, is the longest interval. It spans almost 4 billion years, covering nearly 90% of the geologic history of our Earth. The Cenozoic Era spans 65.5 millions years, making it the shortest era.

2. 2   A drill core, when extracted from the ground, will normally have the oldest layer at the bottom and the youngest at the top. Situated between the Silurian and Carboniferous is the Devonian Period.

3. 4   This fossil is shown at the bottom of the Geologic History chart, letter *M*. Locate letter *M* in the Time Distribution of Fossils section. It is positioned on the Eurypterids bar.

4. 3   Locate these four fossils on the bottom of the Geologic History chart. Using their assigned letters in the Time Distribution of Fossils section, only *L* is positioned in the Triassic Period. Coelophysis existed during this period, and its fossil might be found in Triassic-age sedimentary rocks.

5. 4   Trilobites lived during the Paleozoic era. At the bottom of the Geologic History chart, letters *A*, *B*, and *C* are trilobites. In the Time Distribution of Fossils section, locate the Trilobite bar with these three letters on it. Notice, that this Trilobite bar runs from the Cambrian Period to the Permian Period, spanning all of the Paleozoic Era.

6. 4   On the Eurypterids bar in the Time Distribution of Fossils section, letter *N* corresponds to the diagrammed fossil Stylonurus. Letter *N* is positioned in the Devonian Period.

7. 4   An orogeny is a mountain building process. NYS had four of these episodes, and they are diagrammed (showing mountains) in the Important Geologic Events in NY section. The Alleghenian orogeny is the youngest, occurring in the Pennsylvanian and Permian Periods.

8. 2   In the Important Geologic Events in New York column, go to the Jurassic Period row. Here, it states "Initial opening of Atlantic Ocean." The Jurassic Period is part of the Mesozoic Era.

9. 1   Humans and mammoths are relatively very young compared to most life forms that have existed. At the top of the Life on Earth section, it gives information on humans, mastodonts, and mammoths. This row represents the Pleistocene Epoch.

10. 1   The advance and retreat of the last ice age was a relatively recent geologic event. Go to the Important Geologic Events in New York section. In the first row it states "Advance and retreat of last continental ice." Follow this row to the Epoch column. It lines up with the Pleistocene Epoch.

11. 3 In the Inferred Positions of Earth's Landmasses column, locate the globe labeled 359 mya. This globe shows that NYS is positioned on or very near the equator.

12. 4 Meekoceras fossils are of the Triassic Period. In the Time Distribution of Fossils section, letter *L*, Coelophysis, is positioned in the Triassic Period; therefore, this dinosaur existed during this period.

13. 1 The preserved bedrock representing different time periods in NYS is shown by dark bedrock bars in the NY Rock Record column. Notice that the bedrock of the different periods of the Paleozoic Era are well represented in NYS. The exception is the Permian Period, which has no bar. The bedrock for this time period has either been eroded away or never formed.

14. 4 Locate the fossil Cryptolithus on the bottom of the Geology chart – letter *B*. Locate this letter on the Trilobites bar in the Time Distribution of Fossils column. It is positioned on this bar in the Ordovician Period. Fossil *D*, Valcouroceras, is also positioned in the Ordovician Period. These two fossils, being the same age, might be found together in a sedimentary rock. The other fossils are in different time periods.

15. 4 On the bottom of the Geologic History chart, Valcouroceras is letter *D*. In the Time Distribution of Fossils column, letter *D* is located on the Nautiloids bar within the Ordovician Period.

16. 4 Humans first appeared in the Pleistocene Epoch, which is the present.

17. 4 Locate these four index fossils at the bottom of the Geologic History chart. Using their assigned letters, locate their time of existence in the Distribution of Fossils column.

18.   Answer: Pangaea begins to break up   *or*   Initial opening of Atlantic Ocean
       or North America and Africa separate

   Explanation: The Palisade Sill intrusion is recorded in the Important Geologic Events in NY column. At the end of Triassic Period, it states "Intrusion of Palisades sill." Below this is the statement "Pangaea begins to break up" and above it, "Initial opening of the Atlantic Ocean" is stated.

19.   Answer: Cretaceous   *or*   Late Cretaceous Epoch

   Explanation: Locate the Late Cretaceous Epoch and move to the right stopping at the Important Geologic Events in NY column. Here the question statement is given.

20.   Answer: Tetragraptus

   Explanation: In the Time Distribution of Fossils section, letter *J* is positioned in the lowest part of the Early Ordovician Period, making this fossil the oldest index fossil for this period. On the bottom of the chart is given the name for fossil *J*.

21.   Answer: Asteroid impact

   Explanation: A large asteroid impact would have ejected much dust into the atmosphere causing a climate change, leading to a mass extinction event.

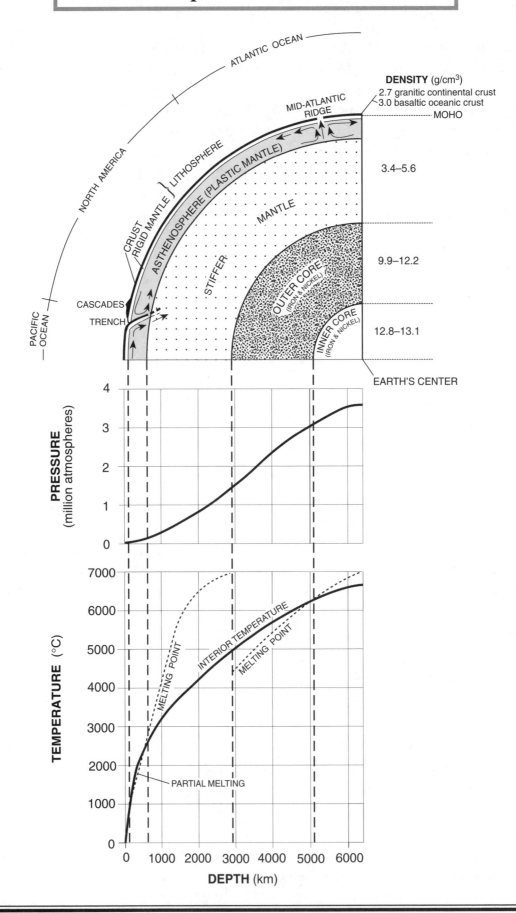

Overview:

If you could travel to the "Center of the Earth," it would be a rough trip. You would enter four different and distinct layers: the solid crust; the mantle, which has a rigid, a plastic, and a larger solid stiffer section; the liquid outer core; and the solid inner core. As you travel downward, the density constantly increases, while the temperature would rise to 6,700°C, and the pressure would be measured in millions of atmospheres. Our understanding of the Earth's interior is mostly based on indirect knowledge because we have never completely penetrated through the crust. This is why the title of the chart contains the word "Inferred" – our best guess. Most of the information on the interior of the Earth has been slowly revealed by the study of seismic waves. These waves are generated by earthquakes and travel deep into the Earth's interior. Seismologists studying seismograms have detected changes in speed, direction, and properties of these waves as they enter different layers. Seismologists know that the seismic *S*-wave cannot pass through liquids. When *S*-waves reach the outer core, they fail to penetrate it; therefore, this core must be a liquid. With more advanced technology, along with creative problem solving, new theories are being proposed and tested, giving better insight in the complexity and dynamics of the Earth's interior.

Earth's Cross-Section Chart:

The Crust – The Earth's outermost layer is the crust. It is relatively cold, thin, and brittle, yet it supplies us with the valuable resources that make life possible. As shown on the chart's upper right side, the crust is layered into the less dense granitic continental crust and the denser basaltic oceanic crust. The continental crust is much thicker than the oceanic crust. Under Density (top right), is the word MOHO. This term represents the boundary zone separating the crust from the rigid mantle.

The Lithosphere – The lithosphere (shown by the two lines) consists of the crust and the uppermost part of the mantle referred to as the rigid mantle. These joined parts collectively make up the plates that are fractured into major and minor plates that slowly but constantly drift. The diagram (lower left) shows the Pacific Ocean Plate colliding with the North American Continental Plate, forming the Cascade Mountain chain. At this convergent zone, a trench is produced as the denser oceanic plate subducts under the overriding less dense continental plate. This area is known as a subduction zone. Shown at the Mid-Atlantic Ridge are hot convection currents, represented by the arrows, breaking through the lithosphere. Here and along other ocean ridges, new magma escapes and quickly solidifies, becoming part of the ocean floor. At these divergent plate boundaries, the youngest ocean floor is found and seafloor spreading is occuring.

The Mantle – Geologist have subdivided the mantle into different sections – the rigid mantle, the asthenosphere, and the stiffer mantle. As already mentioned, the rigid mantle is the lower part of the lithosphere. Under the rigid mantle lies the asthenosphere. The asthenosphere is the soft plastic-like section of the mantle that the plates move over and/or through. Major convection currents are found here, as shown by the arrows in the diagram. These currents generate the energy for plate tectonics. Hot rising convection currents are associated with ocean ridges, and cool sinking convection currents are associated with subduction zones. The mantle is the largest layer of the interior of the Earth in mass and volume. As a whole, the mantle is solid through, in which both the *P* and *S* seismic waves are able to travel.

The Cores – The outer core is a liquid, having a composition of mostly iron and nickel. This liquid core prevents the seismic *S*-wave from traveling through. The inner core, being under so much pressure, remains a solid with an iron-nickel composition. The seismic *P*-wave is transmitted through both cores.

Density – As the depth increases, so does the density. At the upper right side of the chart is the Density section, giving the density range for each layer. Notice that the density values for the two plates are given here.

The Graphs:

Depth axis – Located at the bottom of the graphs is the *x*-axis representing the Depth scale (in km), with 0 being at the surface. The divisions of the Earth's layers are shown by dash lines running down to the Depth axis. Use these lines and this axis to find the distance of each layer from the surface. For example, the outer core starts around 2,900 km and ends around 5,200 km where the inner core starts.

Pressure graph – As the depth increases, so does the pressure. For this graph, pressure is measured in million of atmospheres. The graph line represents the pressure as the depth increases. To find the pressure at a specific layer, locate the dash line for that layer and follow it down, stopping at the intersection of the pressure graph line. Read over to the Pressure axis for the answer. For example, what is the pressure at the start of the outer core? Follow the dash line down at the beginning of the outer core. It intersects the pressure graph line at 1.5 million atmospheres. To find the pressure at a certain depth, use the Depth (km) axis located at the bottom of the page. At the proper depth, move upward to the graph pressure line, then read over to the Pressure axis for the answer. For example, the expected pressure at 5,000 km would be very close to 3 million atmospheres.

Temperature graph – As the depth increases, so does the temperature. The graph line, labeled Interior Temperature, starts near 0°C at the crust and increases to 6,700°C at the Earth's center. Use the same procedure with this graph as with the Pressure graph. For example, what would the temperature be at the boundary between the stiffer mantle and the outer core? Locate this position and follow the dash line downward until it intersects the Interior Temperature graph line. At this position, read over to the Temperature axis. The correct answer is 5,000°C. The inferred Melting Point line is also plotted on this graph shown by small dashes. Within the mantle, the Melting Point line is higher than the Interior Temperature line. For this given condition, the mantle would be a solid. What about the outer core? Here, the Interior Temperature line is higher than the Melting Point line; therefore, the outer core has experienced melting, becoming a liquid. In the inner core, the hottest temperature is found. But, as shown on the graph, the Interior Temperature line is higher than the Melting Point line. This makes the inner core a solid.

Additional Information:

- The speed of seismic waves increases as they enter denser material.

- We have very little direct evidence about the Earth's interior.

- The composition of most meteorites is iron and nickel. Scientists theorize that the Earth's inner core would have a composition similar to meteorites.

- The deepest mine is a diamond/gold mine in South Africa. Its depth is close to 12,800′. At this distance inside the Earth, the temperature of the rocks are close to 300°F (149°C).

- MOHO is named after its discoverer, Andrija Mohorovicic.

- Astronauts left a working seismograph on the Moon. From minor moonquakes (from space impacts), it has been revealed that the moon is completely solid throughout.

Diagrams:

1. **Earth's Interior** – The top of the lithosphere is the Earth's crust. The base of the lithosphere is the rigid part of the mantle. The lithosphere is broken into major and minor sections known as plates. The asthenosphere is situated under the lithosphere and is referred to as the plastic-like mantle. Under the asthenosphere is located the stiffer mantle. The mantle is the largest of all layers of the Earth. The outer core is a liquid, while the inner core is a solid. Both cores have a composition of iron and nickel.

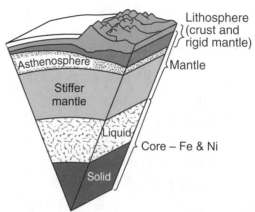

Cross Section of Earth's Interior

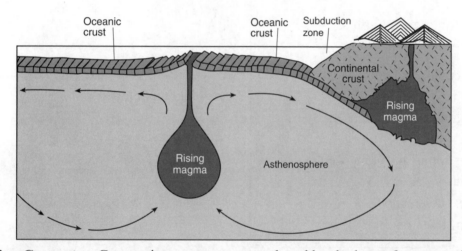

2. **Convection Currents** – Convection currents are produced by the heat of magma within the asthenosphere. Large convection currents circulate within, providing the energy for plate tectonics. When and if the magma breaks through the lithosphere, a rift valley is formed possibly leading to a divergent plate boundary. The convection current eventually cools and sinks deep into the mantle. This area is associated with deep ocean trenches found in subduction zones.

3. **The Dynamic Interior** – The Earth's mantle is very active. Magma plumes rise upward and may warp our crust producing igneous dome mountains. Other magma plumes may produce an active volcanic area known as a mantle hot spot. Here the plume is stationary, but the plates are not. They are constantly but slowly moving over the hot spot. Volcanoes form here, but become extinct when they drift with the plate off the hot spot. Cooler slabs of the lithosphere drift deep into the mantle at the subduction zones where they are melted.

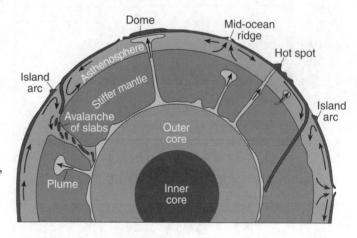

1. Earth's outer core is best inferred to be

   (1) liquid, with an average density
       of approximately 4 g/cm³
   (2) liquid, with an average density
       of approximately 11 g/cm³
   (3) solid, with an average density
       of approximately 4 g/cm³
   (4) solid, with an average density
       of approximately 11 g/cm³   1_____

2. Compared to Earth's crust, Earth's
   inner core is believed to be

   (1) less dense, cooler, and composed
       of more iron
   (2) less dense, hotter, and composed
       of less iron
   (3) more dense, hotter, and composed
       of more iron
   (4) more dense, cooler, and
       composed of less iron   2_____

3. In which layer is the interior temperature
   higher than its melting point?

   (1) crust   (3) outer core
   (2) stiffer mantle (4) inner core   3_____

4. What is the density of the continental crust?

   (1) 3.0 g/cm³   (3) 2.7 g/cm³
   (2) 2.5 g/cm³   (4) 6.2 g/cm³   4_____

5. In which layer of Earth's interior is
   the pressure inferred to be 1.0 million
   atmospheres?

   (1) outer core   (3) rigid mantle
   (2) inner core   (4) stiffer mantle   5_____

6. At which depth below Earth's surface
   is the density most likely 10.5 grams
   per cubic centimeter?

   (1) 1,500 km   (3) 3,500 km
   (2) 2,000 km   (4) 6,000 km   6_____

7. What is the approximate temperature
   at the mantle-outer core boundary?

   (1) 1,500°C   (3) 5,000°C
   (2) 4,500°C   (4) 7,000°C   7_____

Base your answers to question 8 on the
diagram below.

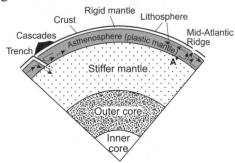

(Not drawn to scale)

8. *a)* The arrows shown in the asthenosphere
   represent the inferred slow circulation of
   the plastic mantle by a process called

   (1) insolation   (3) conduction
   (2) convection   (4) radiation   a_____

   b) The temperature of rock at location *A*
   is approximately

   (1) 600°C   (3) 2,600°C
   (2) 1,000°C   (4) 3,000°C   b_____

   c) Which part of Earth is composed of
   both the crust and the rigid mantle?

   _____

9. Which cross-sectional diagram of a portion of the crust and mantle best shows the pattern of mantle convection currents that are believed to cause the formation of a mid-ocean ridge?

(1)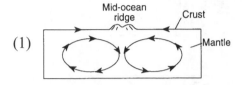

(2)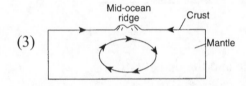

(3)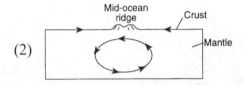

(4)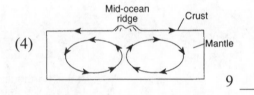

9 _____

10. What is the pressure at the center of the Earth?

(1) 3 millions of atmospheres
(2) 3.2 millions of atmospheres
(3) 3.5 millions of atmospheres
(4) 5,200 millions of atmospheres   10 _____

11. What is the depth to the boundary of the outer core and the inner core?

(1) 5,200 km   (3) 2,900 km
(2) 6,300 km   (4) 5,000 km   11 _____

12. Earth's outer core and inner core are both inferred to be

(1) liquid
(2) solid
(3) composed of a high percentage of iron
(4) under the same pressure   12 _____

13. The diagram shows South America and Africa, the ocean between them, and the ocean ridge and transform faults.
Locations A and D are on the continents.
Locations B and C are on the ocean floor.

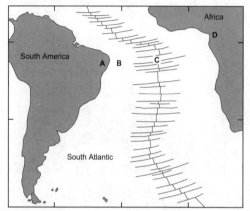

The hottest crustal temperature measurements would most likely be found at location

(1) A   (3) C
(2) B   (4) D   13 _____

14. The cross section below shows a portion of Earth's interior. Layer X is part of Earth's interior.

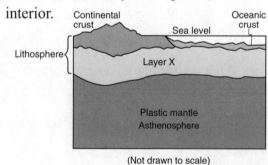

(Not drawn to scale)

a) Identify the part of Earth's lithosphere represented by layer X.

Layer X = _____

b) Which part of the given cross-section would the largest convection currents be located in?

_____

15. Which cross section best represents the relative locations of Earth's asthenosphere, rigid mantle, and stiffer mantle? (The cross sections are not drawn to scale.)

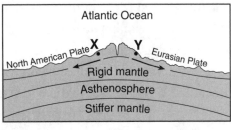

(1)

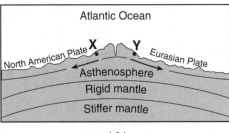

(3)

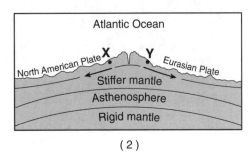

(2)

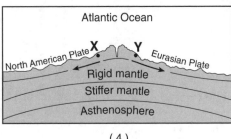

(4)

15 _____

16. *a)* Which data best describe the depth from the Earth's surface and the density of Earth's interior at location *B*?

   (1) Depth: 600 km
       Density: changes from 3.4 g/cm³ to 5.6 g/cm³
   (2) Depth: 1000 km
       Density: averages 4.5 g/cm³
   (3) Depth: 2900 km
       Density: changes from 5.6 g/cm³ to 9.9 g/cm³
   (4) Depth: 5100 km
       Density: averages 11.1 g/cm³

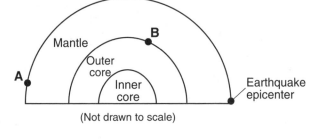

16_____

b) What is the approximate temperature at location *B*? _____

c) What is the approximate pressure at location *B*? _____

17. According to the Earth Science Reference Tables, at what inferred depth is mantle rock partially melted and slowly moving below the lithospheric plates? _____

18. State the density of the oceanic plate. _____ g/cm³

19. In which Earth layer does the pressure reach 3.5 million atmospheres? _____

20. The Cascade Mountain Range was formed by the collision of what two plates?

_____

1. 2   The Density area (right side of chart) shows that 11 g/cm³ is within the range of the outer core. The seismic *S*-wave cannot pass through a liquid. By studying numerous seismograms of major earthquakes, it was determined that *S*-waves were being absorbed by the outer core, making it a liquid.

2. 3   As the depth increases within the Earth, the density and temperature increases. The Earth's Interior chart shows that the composition of the inner core is iron and nickel.

3. 3   Locate the Melting Point and Interior Temperature lines on the Temperature graph within the outer core. The outer core region has a higher interior temperature compared to its melting point. This condition produces a liquid phase.

4. 3   The upper right of the Inferred Properties of Earth's Interior chart gives the density values of the granitic continental crust (2.7 g/cm³) and the basaltic oceanic crust (3.0 g/cm³).

5. 4   Locate the 1.0 million atmospheres position on the Pressure graph. From this position, move directly across until the intersection of the graph line. This position is in the stiffer mantle region.

6. 3   The density of 10.5 g/cm³ fits in the outer core density range as shown by the Inferred Properties of Earth's Interior chart. A depth of 3,500 km is in the outer core.

7. 3   Locate the boundary line of the stiffer mantle and the outer core. At this position, follow the dash line downward until the intersection of the temperature line. The temperature at this position is close to 5,000°C.

8. *a*) 2   Heat is transferred in fluids (liquids and gases) by convection currents. When a fluid becomes warmer, the density decreases and the warmer fluid rises upward being displaced by the cooler fluid, producing a convection current. Conduction is the transfer of heat in solids.

    b) 3   Location *A* corresponds to the top of the stiffer mantle. Follow the dash line downward, representing the top of the stiffer mantle, until the intersection of the temperature line on the temperature graph. The intersection occurs at about 2,600°C.

    c)   Answer: The lithosphere *or* plates

      Explanation: The lithosphere is composed of the crust and the rigid mantle. Together they make up the plates that move over and/or through the asthenosphere. Go to the cross section diagram of the Earth and notice that the lithosphere is shown with two lines, one for the crust and one for the rigid mantle.

Earthquake P-Wave and S-Wave Travel Time

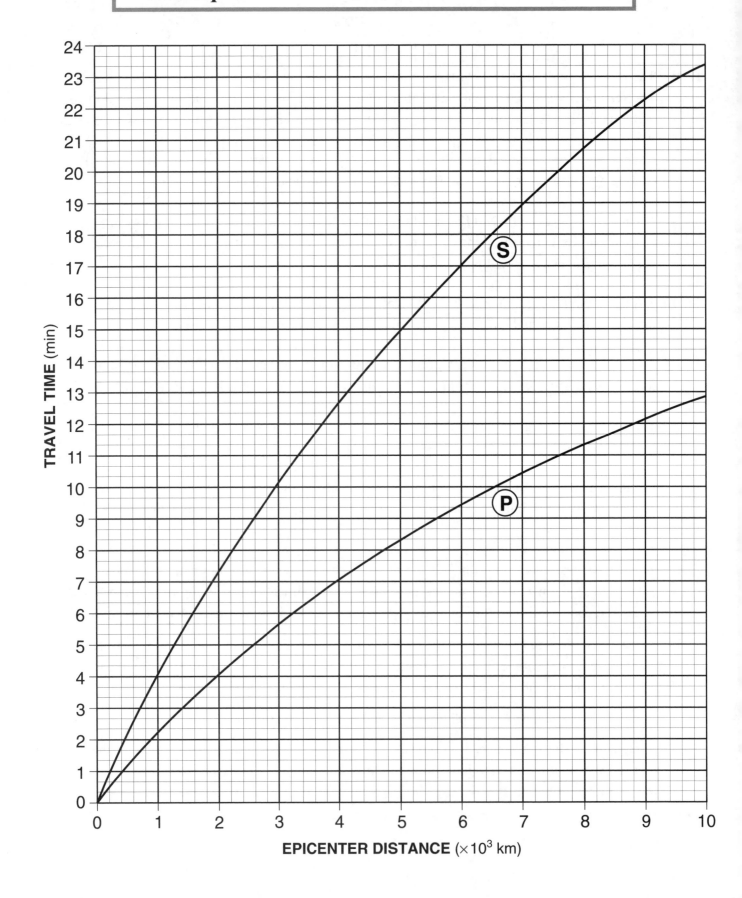

Earthquake P-Wave and S-Wave Travel Time

Earthquake P-Wave and S-Wave Travel Time

Overview:

When the Earth's crust quickly moves and/or snaps, it produces an earthquake, releasing energy in the form of seismic waves that radiate out from the focus. The focus is where the crust broke, while the epicenter is directly above on the Earth's surface. Seismic waves have different properties. The *P*-wave is the fastest wave, reaching seismographs first. Because it can travel through all phases of matter, it can travel completely through the Earth's interior reaching the other side of the Earth. The arrival of the *P*-wave causes little damage, but it is a warning sign that the slower, more destructive *S*-wave is on its way. The *S*-wave causes much destruction due to its shearing action. The *S*-wave can only travel through solids; therefore, it is stopped by the liquid outer core.

Due to the differences in the speed of the *P* and *S*-waves, a separation time occurs between these waves. The farther a seismograph is from the epicenter of an earthquake, the greater the separation time will be for the arrival of the *P* and *S*-wave. Using the known speeds of the *P* and *S*-waves (as shown on the graph on the previous page) and the arrival time of these waves as recorded by a seismograph, the distance to the epicenter can determined. Using the recordings (seismograms) from three seismographs, the location of the epicenter can be determined.

The Graph:

The Axis – The *x*-axis is the Epicenter Distance scale. The bold dark vertical lines represent increments of 1,000 km (1×10^3 km). The lighter lines represent intervals of 200 km. The *y*-axis is the Travel Time scale. The bold dark lines represent intervals of a minute, while the lighter lines represent 20 seconds.

The Graph – The bold graph lines represent the speed of the *P* and *S* seismic waves. When an earthquake occurs, the *P* and *S*-waves are generated at the same time. On the graph this would be at minute 0. These waves immediately start to separate because the *P*-wave is faster. Thus, the two lines become farther apart as distance increases, producing a separation time. To get the travel time for either wave for a certain distance, move upward from the given distance on the Epicenter Distance axis until the intersection of the correct seismic line. At this intersection point, read over to the Travel Time axis. For example, a *P*-wave traveling 7,200 km would take a travel time of 00:10:40 (10 min 40 sec). The slower *S*-wave would take 00:19:20 (19 min 20 sec) to go the same distance. Given the travel time of a seismic wave, the travel distance can be determined. For example, how far would a *P*-wave travel in 8:20? Locate this time on the Travel Time axis and move across to the *P*-wave line. At the intersection point, move directly down to the Epicenter Distance axis. The answer is 5,000 km. For the same travel time, the slower *S*-wave would have only traveled 2,400 km. So as you can see, these problems are "up and over" or "over and down" problems.

Epicenter Distance – If you know the arrival time of both waves, the distance to the epicenter can be determined. The arrival times of the *P*-wave and the *S*-wave may be given to you, or you may have to interpret them from a seismogram. Once you have determined the arrival time of both waves, subtract the *P* arrival time from the *S* arrival time. This is the separation time. Next, position the edge of a piece of paper on the Travel Time axis making a mark at the 0 time and another mark at the separation time. Take this paper with these marks to the graph, position it until the marks fit vertically between the *P* and *S* lines. Reading directly down from this "fitted" position to the *x*-axis will give you the epicenter distance.

Example 1: If the *P*-wave arrived at 10:20:10 and the *S*-wave arrived at 10:26:30, how far away is the epicenter?

Solution: Subtracting the *P* and *S*-waves, the separation time is 00:6:20. Using a piece of paper, mark off this time interval from the Travel Time axis. Move this paper, with the two marks, until these marks fit vertically between the *P* and *S* graph lines. Read directly down to the Epicenter Distance axis. If done correctly, the epicenter distance will be 4,800 km ($\pm$200 km).

Example 2:

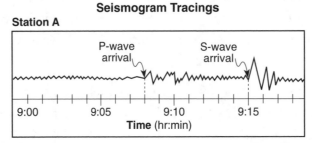

Seismogram Tracings

Station A

P-wave arrival   S-wave arrival

9:00   9:05   9:10   9:15

Time (hr:min)

From the above seismogram, what is the distance from Station A to the epicenter?

Solution: The separation time of the *P* and *S*-wave as shown on the seismogram is 7 min 00 sec (00:07:00). This time interval, marked off from the Travel Time axis, fits vertically between the *P* and *S*-wave graph lines at a distance of 5,600 km ($\pm$200 km).

Epicenter Location – The location of the epicenter is determined by using the epicenter distances from three seismic stations. By drawing three circles on a map with the radius of each representing the distances to the epicenter, there will be a common intersection point of the three circles. This intersection point is the location of the earthquake's epicenter (see diagram 3, page 131).

Origin Time – The origin time of an earthquake is the exact time that the earthquake occurred. The origin time can be determined by knowing the travel time of the *P*-wave to a seismic station. This time is then subtracted from the time that the seismograph recorded the arrival of the *P*-wave.

Example: A seismograph recorded the first *P*-wave at 10:10:40 and it was determined that the *P*-wave traveled 2,200 km. What was the origin time of the earthquake?

Solution: It takes 00:04:20 (4 min 20 sec) for a *P*-wave to travel 2,200 km. Subtract this time from 10:10:40, the origin time would be 10:06:20.

Additional Information:

- By studying the amplitude of the seismic waves as recorded on the seismogram, the magnitude or strength of the earthquake can be determined. This becomes its Richter Scale number. The Mercalli Scale is used for measuring the intensity of an earthquake using observable damage.

- A tsunami is a seismic sea wave. It can be generated by an underwater earthquake, landslide or volcanic eruption.

- During an earthquake, there is a major area that surrounds the Earth that does not receive any *P* or *S*-waves. This area is known as the shadow zone and is caused by the refraction, reflection, and absorption of seismic waves as they encountered different density layers of our Earth.

Diagrams:

1. **Seismic Waves Properties** – When an earthquake occurs, seismic waves radiate away from the epicenter, traveling on the Earth's surface and deep into the interior. As the *P* and *S*-waves enter new layers of the Earth's interior they undergo reflection, refraction, absorption and change of speed. A seismograph located at position *X* will record both the *P* and *S*-waves. The liquid outer core will absorbed the *S*-waves and no further travel occurs. Thus, only the *P*-waves will be recorded by seismograph *Z*, for these waves can travel through all phases of matter. The gray area is known as the shadow zone, where no seismic waves are recorded (location *Y* and *W*). This occurs due to the refraction (bending) of waves away from this area by the liquid outer core.

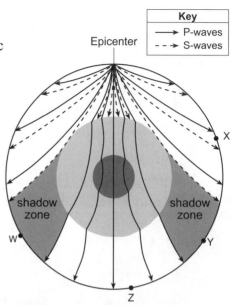

2. **Seismogram Reading** – The *P*-wave is recorded first on a seismogram, then the slower more destructive *S*-wave arrives. The *S*-wave's magnitude (strength) can be determined by the height of the recorded wave. This seismogram shows a separation time of 00:05:40 (5 min and 40 sec). Marking off this time interval on the Travel Time axis, and fitting this distance vertically between the *S* and *P* graph lines gives the epicenter distance. For this earthquake, the epicenter distance is 4,000 km.

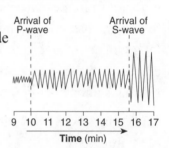

3. **Epicenter Location** – To get the epicenter location, the recordings from three seismograph stations are needed. From each station, the distance to the epicenter needs to be determined using the separation time of the *P* and *S*-waves. When three circles are drawn representing the three epicenter distances, the common intersection point is the location of the earthquake.

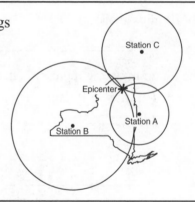

4. **The Mercalli Scale** – This scale measures the intensity of an earthquake and is based on observable damage to the surroundings.

I	Instrumental: detected only by instruments	VII	Very strong: noticed by people in autos Damage to poor construction
II	Very feeble: noticed only by people at rest	VIII	Destructive: chimneys fall, much damage in substantial buildings, heavy furniture overturned
III	Slight: felt by people at rest Like passing of a truck	IX	Ruinous: great damage to substantial structures Ground cracked, pipes broken
IV	Moderate: generally perceptible by people in motion Loose objects disturbed	X	Disastrous: many buildings destroyed
V	Rather strong: dishes broken, bells rung, pendulum clocks stopped People awakened	XI	Very disastrous: few structures left standing
VI	Strong: felt by all, some people frightened Damage slight, some plaster cracked	XII	Catastrophic: total destruction

1. How long would it take for the first S-wave to arrive at a seismic station 4,000 kilometers away from the epicenter of an earthquake?

   (1) 5 min 40 sec
   (2) 7 min 0 sec
   (3) 12 min 40 sec
   (4) 13 min 20 sec   1 _____

2. A P-wave takes 8 minutes and 20 seconds to travel from the epicenter of an earthquake to a seismic station. Approximately how long will an S-wave take to travel from the epicenter of the same earthquake to this seismic station?

   (1) 6 min 40 sec
   (2) 9 min 40 sec
   (3) 15 min 00 sec
   (4) 19 min 00 sec   2 _____

3. How far will a S-wave travel in 10 minutes and 40 seconds?

   (1) 3,200 km   (3) 5,600 km
   (2) 3,900 km   (4) 7,200 km   3 _____

4. An earthquake's P-wave arrived at a seismograph station at 02 hours 40 minutes 00 seconds. The earthquake's S-wave arrived at the same station 2 minutes later. What is the approximate distance from the seismograph station to the epicenter of the earthquake?

   (1) 1,100 km   (3) 3,100 km
   (2) 2,400 km   (4) 4,000 km   4 _____

5. A seismograph recorded the first P-wave at 00:09:20. This P-wave traveled 2,200 km. What was the origin time of the earthquake?

   (1) 00:14:00   (3) 00:01:20
   (2) 00:05:00   (4) 00:10:20   5 _____

6. The diagram below represents three seismograms showing the same earthquake as it was recorded at three different seismic stations, A, B, and C.

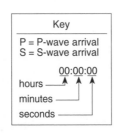

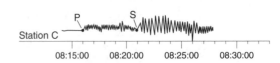

   Which statement correctly describes the distance between the earthquake epicenter and these seismic stations?

   (1) A is closest to the epicenter, and C is farthest from the epicenter.
   (2) B is closest to the epicenter, and C is farthest from the epicenter.
   (3) C is closest to the epicenter, and A is farthest from the epicenter.
   (4) A is closest to the epicenter, and B is farthest from the epicenter.   6 _____

Base your answers to question 7 on the diagram and map below. The diagram shows three seismograms of the same earthquake recorded at three different seismic stations, *X*, *Y*, and *Z*. The distances from each seismic station to the earthquake epicenter have been drawn on the map. A coordinate system has been placed on the map to describe locations. The map scale has not been included.

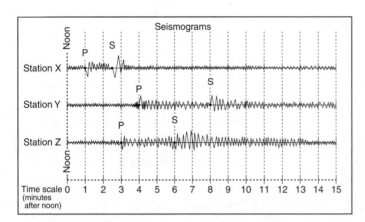

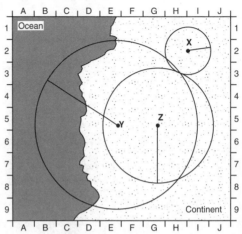

7. *a)* Approximately how far away from station *Z* is the epicenter?

   (1) 1,300 km
   (2) 1,800 km
   (3) 3,900 km
   (4) 5,200 km   a _____

   b) The *S*-waves from this earthquake that travel toward Earth's center will

   (1) be deflected by Earth's magnetic field
   (2) be totally reflected off the crust-mantle interface
   (3) be absorbed by the liquid outer core
   (4) reach the other side of Earth faster than those that travel around Earth in the crust   b _____

   c) Seismic station *X* is 800 kilometers from the epicenter. Approximately how long did it take the *P*-wave to travel to station *X*?

   (1) 1 min 50 sec
   (2) 2 min 50 sec
   (3) 3 min 20 sec
   (4) 6 min 30 sec   c _____

   d) On the map, which location is closest to the epicenter of the earthquake?

   (1) E–5   (3) H–3
   (2) G–1   (4) H–8   d _____

8. A seismic station in Massena, New York, recorded the arrival of the first *P*-wave at 1:30:00 (1 hour, 30 minutes, 00 seconds) and the first *S*-wave from the same earthquake at 1:34:30.

   a) Determine the distance, in kilometers, from Massena to the epicenter of this earthquake. _____ km

   b) State what additional information is needed to determine the location of the epicenter of this earthquake.

9. Scientists have inferred the structure of Earth's interior mainly by analyzing

    (1) the Moon's interior
    (2) the Moon's composition
    (3) Earth's surface features
    (4) Earth's seismic data   9 _____

Base your answers to question 10 on the earthquake seismogram below.

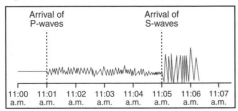

10. *a)* Approximately how far away is the epicenter?

    (1) 2,000 km
    (2) 2,600 km
    (3) 3,500 km
    (4) 4,400 km   a _____

    b) How many additional seismic stations must report seismogram information in order to locate the epicenter?

    (1) one   (3) three
    (2) two   (4) four   b _____

11. The distance from an epicenter of an earthquake to New York City is 3,000 kilometers. What was the approximate travel time for the *P*-waves from this epicenter to New York City?

    (1) 1 min 20 sec
    (2) 5 min 40 sec
    (3) 7 min 30 sec
    (4) 10 min 00 sec   11 _____

12. A huge undersea earthquake off the Alaskan coastline could produce a

    (1) tsunami   (3) hurricane
    (2) cyclone   (4) thunderstorm   12 _____

13. An earthquake's first *P*-wave arrives at a seismic station at 12:00:00. This *P*-wave has traveled 6,000 kilometers from the epicenter. At what time will the first *S*-wave from the same earthquake arrive at the seismic station?

    (1) 11:52:20   (3) 12:09:20
    (2) 12:07:40   (4) 12:17:00   13 _____

14. The cutaway diagram below shows the paths of earthquake waves generated at point *X*.

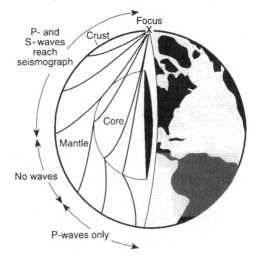

Only *P*-waves reach the side of Earth that is opposite the focus because *P*-waves

    (1) are stronger than *S*-waves
    (2) travel faster than *S*-waves
    (3) bend more than *S*-waves
    (4) can travel through liquids and *S*-waves cannot   14 _____

Base your answers to question 15 on the map below. The map shows point X, which is the location of an earthquake epicenter, and point A, which is the location of a seismic station.

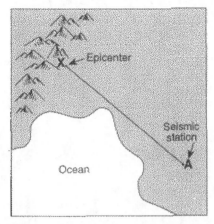

Scale

Distance (× 10³ km)

15. *a*) Approximately how long did the earthquake's *P*-wave take to arrive at the seismic station?

(1) 3 min 40 sec
(2) 5 min 10 sec
(3) 6 min 20 sec
(4) 11 min 5 sec   a _____

b) Which statement best describes the arrival of the initial *S*-wave at the seismic station?

(1) It arrived later than the *P*-wave because *S*-waves travel more slowly.
(2) It arrived earlier than the *P*-wave because *S*-waves travel faster.
(3) It arrived at the same time as the *P*-wave because *S*-waves and *P*-waves have the same velocity on Earth's surface.
(4) It never reached location *A* because *S*-waves can travel only through a liquid medium.   b _____

16. An earthquake occurs at 12:02 p.m. A seismic station records the first S-wave at 12:19 p.m. Which set of data shows the approximate arrival time of the first P-wave and the distance to the epicenter?

(1) 12:11:25 p.m. and 4000 km
(2) 12:11:25 p.m. and 6000 km
(3) 12:19:40 p.m. and 4000 km
(4) 12:19:40 p.m. and 6000 km   16 _____

17. How far from an earthquake epicenter is a city where the difference between the *P*-wave and *S*-wave arrival times is 10 minutes and 30 seconds?

(1) 1.7×10^3 km   (3) 3.5×10^3 km
(2) 9.9×10^3 km   (4) 4.7×10^3 km

17 _____

18. The seismogram below shows *P*-wave and *S*-wave arrival times at a seismic station following an earthquake.

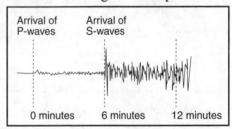

The distance to the epicenter is approximately

(1) 1,600 km   (3) 4,400 km
(2) 3,200 km   (4) 5,600 km   18 _____

19. A seismograph recorded the first *P*-wave at 00:08:40. It was determined that the *P*-wave traveled 1,200 km. What was the origin time of the earthquake?

(1) 00:11:20   (3) 00:07:40
(2) 00:10:40   (4) 00:06:00   19 _____

20. The map shows the location of an earthquake epicenter in New York State. Seismic stations *A*, *B*, and *C* received the data used to locate the earthquake epicenter.

The seismogram recorded at station *A* would show the

(1) arrival of *P*-waves, only
(2) earliest arrival time of *P*-waves
(3) greatest difference in the arrival times of *P*-waves and *S*-waves
(4) arrival of *S*-waves before the arrival of *P*-waves

20 _____

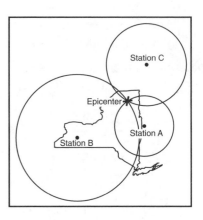

21. An earthquake's magnitude can be determined by

(1) analyzing the seismic waves recorded by a seismograph
(2) calculating the depth of the earthquake faulting
(3) calculating the time the earthquake occurred
(4) comparing the speed of *P*-waves and *S*-waves

21_____

22. List two actions a homeowner could take to prepare the home or family for an earthquake.

1_____

2_____

Seismogram Tracings

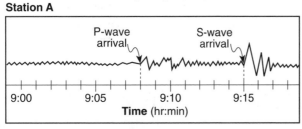

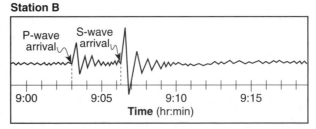

23. *a)* From the diagrams above, explain how the seismic tracings recorded at station *A* and station *B* indicate that station *A* is farther from the earthquake epicenter than station *B*.

_____

b) Seismic station *A* is located 5,400 kilometers from the epicenter of the earthquake. How much time would it take for the first *S*-wave produced by this earthquake to reach seismic station *A*?

_____

c) Why is the magnitude of the *S*-wave larger at station *B*? _____

_____

24. State one safety measure that the state of Hawaii has or should implement to prevent loss of life from a future tsunami.

_____

1. 3 Locate 4,000 km on the Epicenter Distance scale. From this position, move directly up until the intersection of the *S*-wave line. At this intersection point, read over to the Travel Time axis.

2. 3 First find out how far a *P*-wave travels in 8:20. Locate this given time on the Travel Time axis. From this time, move directly across to the intersection of the *P*-wave line. Reading down from this point, the distance to the epicenter is 5,000 km. Using this distance, move straight up to the intersection of the *S*-wave line. This intersection point is 15 minutes as read on the Travel Time axis.

3. 1 Go to the 10:40 line on the Travel Time axis. Move directly across until the intersection of the *S*-wave line. Read down from this intersection position to the Epicenter Distance axis.

4. 1 The separation time between the *P* and *S*-wave is 2 minutes. Take a piece of paper and place it along the Travel Time axis. Mark off a 2 minute interval from this axis. Move this paper over to the graph, keeping it vertical, until the two marks fit between the *P* and *S* graph lines. Read down from this fitted position to the Epicenter Distance axis.

5. 2 It takes 4 minutes and 20 seconds (00:04:20) for a *P*-wave to travel 2,200 km. The origin time of the earthquake would have occurred 00:04:20 before the recorded time, or at 00:05:00.

6. 3 The closer a seismic station is to the epicenter of an earthquake, the shorter the separation time between the *P* and *S*-waves will be. The separation times are: *A* = 9 min, *B* = 7 min, and *C* = 5 min. Station *C*, having the smallest separation time, must be the closest station and Station *A* must be the farthest.

7. *a*) 2 The seismogram for Station *Z* shows a 3 minute separation time between the arrival of the *P*-wave and the arrival of the *S*-wave. Take a piece of paper and place it along the Travel Time axis. Mark off a 3 minute interval from this axis. Move this paper over to the graph, keeping it vertical, until the two marks fit between the *P* and *S* graph lines. Read down from this fitted position to the Epicenter Distance axis.

b) 3 The *S*-waves cannot be transmitted through liquids. When these waves reach the outer core, they are stopped, being absorbed by the liquid outer core.

c) 1 Find this distance, 800 km, on the Epicenter Distance scale. From this position, move upward to the intersection of the graph *P*-wave line, then read over to the Travel Time axis.

d) 3 The location of the epicenter of an earthquake is found by a common intersection point of three circles, representing the correct distance to the epicenter from each seismic station. On the map, all three circles have a common intersection at location H-3.

8. *a*) Answer: 3,000 km (±200 km)

Explanation: The separation time is 4:30. Mark off this time interval from the Travel Time axis. It fits vertically between the *P* and *S* lines at a distance of 3,000 km.

b) Answer: Data from two additional seismic stations.

Explanation: Three seismic stations are needed to determine the location of the epicenter.

Dewpoint (°C)

Dry-Bulb Temperature (°C)	Difference Between Wet-Bulb and Dry-Bulb Temperatures (C°)															
	0	1	2	3	4	5	6	7	8	9	10	11	12	13	14	15
−20	−20	−33														
−18	−18	−28														
−16	−16	−24														
−14	−14	−21	−36													
−12	−12	−18	−28													
−10	−10	−14	−22													
−8	−8	−12	−18	−29												
−6	−6	−10	−14	−22												
−4	−4	−7	−12	−17	−29											
−2	−2	−5	−8	−13	−20											
0	0	−3	−6	−9	−15	−24										
2	2	−1	−3	−6	−11	−17										
4	4	1	−1	−4	−7	−11	−19									
6	6	4	1	−1	−4	−7	−13	−21								
8	8	6	3	1	−2	−5	−9	−14								
10	10	8	6	4	1	−2	−5	−9	−14	−28						
12	12	10	8	6	4	1	−2	−5	−9	−16						
14	14	12	11	9	6	4	1	−2	−5	−10	−17					
16	16	14	13	11	9	7	4	1	−1	−6	−10	−17				
18	18	16	15	13	11	9	7	4	2	−2	−5	−10	−19			
20	20	19	17	15	14	12	10	7	4	2	−2	−5	−10	−19		
22	22	21	19	17	16	14	12	10	8	5	3	−1	−5	−10	−19	
24	24	23	21	20	18	16	14	12	10	8	6	2	−1	−5	−10	−18
26	26	25	23	22	20	18	17	15	13	11	9	6	3	0	−4	−9
28	28	27	25	24	22	21	19	17	16	14	11	9	7	4	1	−3
30	30	29	27	26	24	23	21	19	18	16	14	12	10	8	5	1

Overview:

Dewpoint indicate the amount of moisture in the air. The higher the dewpoint, the higher the moisture content of the air. The dewpoint temperature is the temperature the air would need to be cooled down to in order for the water vapor within the air to condense, changing it from a gas to a liquid. When this happens, the air is saturated and clouds, fog, or dew starts forming. Normally, the dewpoint temperature is lower than the air temperature (it will never be higher), and the air is not saturated. If the dewpoint and air temperature become closer, the relative humidity increases. When these two temperatures are the same, the air is saturated and the relative humidity is 100%. As one can see, when the dewpoint temperature is close to the air temperature, relative humidity is high, and precipitation may be in the forecast.

To measure the dewpoint, an instrument called a sling psychrometer is used. This instrument has two thermometers, the dry-bulb thermometer and the wet-bulb thermometer. The dry-bulb thermometer records the air temperature. The wet-bulb thermometer contains a moistened wick around its bulb. When swung through the air, the moisture on the wet bulb evaporates, which removes heat, causing a lowering of the wet-bulb temperature. This gives a difference in temperature readings of the two thermometers. Using this difference of temperatures and the Dewpoint chart, the dewpoint temperature of the air can be determined.

The Chart:

As mentioned, the sling psychrometer will have two thermometers, the dry-bulb thermometer and the wet-bulb thermometer. Subtracting these two temperature readings gives the temperature difference. At the top of the chart, locate the correct column representing the difference in temperatures. Move down in this column until you reach the correct Dry-Bulb Temperature row. This intersection position is your dewpoint temperature. Let's try one. What is the dewpoint temperature if the dry-bulb temperature is 20°C and the wet-bulb temperature is 16°C? Subtracting these temperatures gives the difference of 4°C. Locate this column at the top of the chart and move downward until the intersection of the Dry-Bulb Temperature of 20°C. The given dewpoint is 14°C; therefore, if this air which is at 20°C cools down to this temperature (14°C), condensation will occur, and produce visible moisture droplets as found in a cloud.

Additional Information:

- On a clear night, the atmosphere cools quickly hitting its dewpoint temperature. This produces dew forming on the Earth's surface, or possibly produce "atmospheric dew," called fog.

- Condensation nuclei are microscopic airborne particles of smoke, dust, salt, etc. These particles need to be present for condensation to occur.

Diagrams:

1. **Sling Psychrometer** – The sling psychrometer is used to get the dewpoint temperature as well as the relative humidity. When swung, 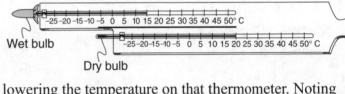 the water on the wet bulb evaporates lowering the temperature on that thermometer. Noting the difference between the two temperatures and using the proper chart will give you the dewpoint temperature. When the air is saturated the two temperatures will be the same and the relative humidity is 100%.

2. **Dewpoint and Air Temperature Relationship** – The table shows the air temperature and dewpoint at four locations. Based on these measurements, location *B* would have the greatest chance of precipitation. This is because the dewpoint temperature, equals the air temperature. At this condition, the air is saturated with water vapor and condensation begins.

Dewpoint Table

Location	A	B	C	D
Air temperature (°F)	80	60	45	35
Dewpoint (°F)	60	60	35	25

3. **Fog** – When warm, moist air moves over the cold land surface, the air temperature drops. If this temperature reaches the dewpoint temperature, condensation occurs. As shown in this diagram, the condensing water vapor has produced fog. Fog is a thick cloud of water droplets suspended in the air near the Earth's surface.

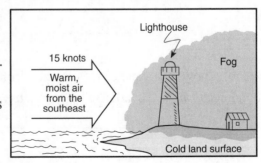

1. What is the dewpoint when the dry-bulb temperature is 14°C and the wet-bulb temperature is 8°C?

   (1) 1°C   (3) 6°C
   (2) –9°C   (4) 22°C   1 _____

2. What is the dewpoint when the dry-bulb temperature is 16°C and the wet-bulb temperature is 12°C?

   (1) –19°C   (3) 7°C
   (2) –16°C   (4) 9°C   2 _____

3. A student used a sling psychrometer to measure the dewpoint of the air. If the dewpoint was 6°C and the dry-bulb temperature was 10°C, what was the wet-bulb temperature?

   (1) 5°C   (3) 8°C
   (2) 7°C   (4) 10°C   3 _____

4. The station model shows several weather variables recorded at a particular location. (see page 158)

   What was the most likely dewpoint at this location?
   (1) 32°F   (3) 62°F
   (2) 40°F   (4) 70°F   4 _____

5. A student uses a sling psychrometer outdoors on a clear day. The dry-bulb (air) temperature is 10°C. The water on the wet bulb will most likely

   (1) condense, causing the wet-bulb temperature to be higher than the air temperature.
   (2) condense, causing the wet-bulb temperature to be equal to the air temperature.
   (3) evaporate, causing the wet-bulb temperature to be lower than the air temperature.
   (4) evaporate, causing the wet-bulb temperature to be equal to the air temperature.   5 _____

6. Which weather change usually occurs when the difference between the air temperature and the dewpoint temperature is decreasing?

   (1) The amount of cloud cover decreases.
   (2) The probability of precipitation decreases.
   (3) The relative humidity increases.
   (4) The barometric pressure increases.
        6 _____

7. What is the difference between the dry-bulb temperature and wet-bulb temperature when the dewpoint is 9°C and the dry bulb is 18°C?

   (1) 1°C   (3) 4°C
   (2) 3°C   (4) 5°C   7 _____

8. Describe how hurricane clouds formed from water vapor. Include the terms "dewpoint" and either "condensation" or "condense" in your answer.

_____

9. What is the dewpoint temperature when the dry-bulb temperature is 12°C and the wet-bulb temperature is 7°C?

(1) 1°C   (3) 6°C

(2) 5°C   (4) 4°C   9 _____

10. What is the dewpoint when the dry-bulb temperature is 24°C and the wet-bulb temperature is 15°C?

(1) 8°C   (3) 36°C

(2) –18°C   (4) 4°C   10 _____

Note: Question 11 has only three choices.

11. Weather-station measurements indicate that the dewpoint temperature and air temperature are getting farther apart over the past 3 hours. The chance of precipitation during this time span is

(1) increasing
(2) decreasing
(3) remains the same   11 _____

12. State the phase change that occurs at the dewpoint. _____

13. Which graph best shows the relationship between the probability of precipitation and the difference between air temperature and dewpoint?

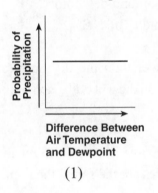

(1)

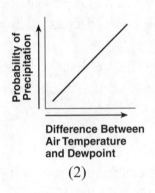

(2)

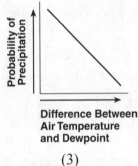

(3)

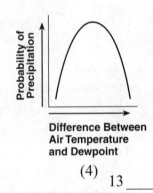

(4)

13 _____

The following weather data was collected at Boonville, New York.

14. Describe what will occur when the air temperature drops one more degree.

Air temperature	65°F
Dewpoint	64°F
Visibility	2 miles
Present weather	--------
Wind direction	from the west
Wind speed	5 knots
Amount of cloud cover	100%
Barometric pressure	996.2 millibars

_____

15. Name the instrument that is used to obtain the dewpoint temperature.

_____

1. 1   The difference between the dry-bulb and the wet-bulb temperature is 6°C. Locate this temperature at the top of the Dewpoint chart. Move down in this column until you reached the Dry-Bulb Temperature row of 14°C. At this intersection point, the dewpoint temperature is given as 1°C.

2. 4   The difference between the dry-bulb and the wet-bulb temperature is 4°C. Go to this temperature at the top of the Dewpoint chart. Move down in this column until you reached the Dry-Bulb Temperature row of 16°C. At this intersection point, the dewpoint temperature is given as 9°C.

3. 3   Go down the Dry-Bulb Temperature column stopping at the given dry-bulb temperature of 10°C. Move across until you reach the dewpoint of 6°C. At this position, this column represents a 2-degree difference between the wet-bulb and dry-bulb temperatures. Because the wet-bulb temperature will be lower than the dry-bulb temperature, subtracting 2 degrees from the dry-bulb temperature gives the answer of 8°C (10°C – 2°C = 8°C).

4. 3   The station model has the symbol for rain (•). When rain is present, the air and dewpoint temperatures will be the same. These conditions produce a relative humidity of 100%.

5. 3   On a clear day, the air is unsaturated causing the wet-bulb to be lower than the dry-bulb (air) temperature. This happens due to the evaporation of some of the moisture on the wet bulb. Remember, evaporation is a cooling process.

6. 3   As the dewpoint temperature and air temperature (the dry bulb) become closer (difference is decreasing), the relative humidity increases. When these two temperatures are the same, the relative humidity is 100%, and the air is saturated and condensation begins.

7. 4   Go down the Dry-Bulb Temperature column stopping at 18°C. Move across in this row stopping at the dewpoint of 9°C. This intersection position is in the 5°C column.

8.   Answer: As water vapor enters the air, or the air temperature drops, eventually the air will become saturated. Now the dewpoint temperature is the same as the air temperature (RH = 100%) and condensation occurs producing clouds.

   Note: This is the basic process for all clouds.

Dry-Bulb Tempera-ture (°C)	Difference Between Wet-Bulb and Dry-Bulb Temperatures (C°)															
	0	1	2	3	4	5	6	7	8	9	10	11	12	13	14	15
−20	100	28														
−18	100	40														
−16	100	48														
−14	100	55	11													
−12	100	61	23													
−10	100	66	33													
−8	100	71	41	13												
−6	100	73	48	20												
−4	100	77	54	32	11											
−2	100	79	58	37	20	1										
0	100	81	63	45	28	11										
2	100	83	67	51	36	20	6									
4	100	85	70	56	42	27	14									
6	100	86	72	59	46	35	22	10								
8	100	87	74	62	51	39	28	17	6							
10	100	88	76	65	54	43	33	24	13	4						
12	100	88	78	67	57	48	38	28	19	10	2					
14	100	89	79	69	60	50	41	33	25	16	8	1				
16	100	90	80	71	62	54	45	37	29	21	14	7	1			
18	100	91	81	72	64	56	48	40	33	26	19	12	6			
20	100	91	82	74	66	58	51	44	36	30	23	17	11	5		
22	100	92	83	75	68	60	53	46	40	33	27	21	15	10	4	
24	100	92	84	76	69	62	55	49	42	36	30	25	20	14	9	4
26	100	92	85	77	70	64	57	51	45	39	34	28	23	18	13	9
28	100	93	86	78	71	65	59	53	47	42	36	31	26	21	17	12
30	100	93	86	79	72	66	61	55	49	44	39	34	29	25	20	16

Overview:

Relative humidity (RH) is the percentage of water vapor in the air relative to how much the air can hold (is saturated) at the current temperature. Both relative humidity and dewpoint play an important role on our moisture comfort level. High temperatures with high dewpoint and RH values produce that hot, sticky, oppressive weather. Cold temperatures along with dry air produce low relative humidity readings, which can cause dry skin condition. From day to day (even hour to hour), the relative humidity is constantly changing due to changes in air temperature, moisture amount, and the dewpoint temperature. Overall, when a low (L) pressure system is advancing and the relative humidity is heading toward 100%, there is a good chance of precipitation.

Relative humidity is measured by using a sling psychrometer, the same instrument used to measure the dewpoint temperature. This instrument has two thermometers, a dry-bulb thermometer that measures the air temperature and a wet-bulb thermometer. As the instrument is swung, the moisture on the wet bulb evaporates, removing heat from the thermometer bulb, causing the temperature to decrease. Obtaining the difference between the dry-bulb temperature and the wet-bulb temperature readings and using the RH chart, one can arrive at the relative humidity. But remember, this weather variable is constantly changing, just as weather does.

The Chart:

This chart has the same layout as the Dewpoint chart. From a sling psychrometer, get the air temperature by reading the dry-bulb thermometer, and then get the lower wet-bulb temperature. Subtracting these two temperatures gives the number (in degrees) to use in the top section labeled "Difference Between Wet-Bulb and Dry-Bulb Temperatures." Using this difference number and its corresponding column, move down until you reach the correct Dry-Bulb Temperature row. At the intersection of these two numbers will be the RH value. For example, what is the RH if the dry-bulb temperature is 14°C and the wet-bulb temperature is 8°C? Solution: The difference between the dry-bulb and the wet-bulb temperatures is 6°C. Go to this column at the top of the RH chart. Staying in this column, move down stopping at the Dry-Bulb Temperature row of 14°C. At the intersection position, the answer shown is 41% RH.

Additional Information:

- Conditions that are favorable for rapid evaporation are: low RH, high temperature, and windy conditions.

Diagrams:

1. **Cloud Formation** – When rising air hits the dewpoint temperature, water vapor condenses producing cloud droplets. Billions of these tiny floating droplets produce a cloud. When many of these droplets join and become large and heavy enough to fall to the Earth, precipitation is produced. Precipitation cleans the atmosphere by removing suspended particles.

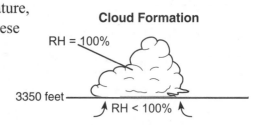

2. **Station Models and Relative Humidity** – At noon the air temperature was 28°F and the dewpoint was 26°F. From these readings, the relative humidity was close to 80%. Eight hours later these two temperatures were the same, making the relative humidity 100%. The air was saturated with water vapor, condensation occurred, and precipitation was in the form of snow.

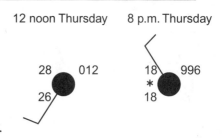

3.

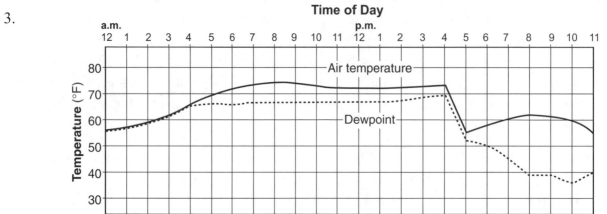

Recording of Two Weather Variables – From 12 a.m. to 4 a.m., the dewpoint and air temperatures were the same making the RH 100% and rain occurred. At 4 p.m., a cold front passed through, and thunderstorms were reported. With the passage of the cold front (5 p.m.), the relative humidity decreased as shown by the separation of the dewpoint and air temperatures.

1. A parcel of air has a dry-bulb temperature reading of 16°C and a wet-bulb temperature reading of 13°C. What is the relative humidity?

   (1) 11%   (3) 71%
   (2) 13%   (4) 80%   1 _____

2. What is the relative humidity when the air temperature is 29°C and the wet-bulb temperature is 23°C?

   (1) 6%   (3) 54%
   (2) 20%   (4) 60%   2 _____

3. If air has a dry-bulb temperature of 2°C and a wet-bulb temperature of –2°C, what is the relative humidity?

   (1) 11%   (3) 36%
   (2) 20%   (4) 67%   3 _____

4. A parcel of air has a dry-bulb temperature of 18°C and a wet-bulb temperature of 10°C. What are the dewpoint and the relative humidity?

   (1) 5°C and 19%
   (2) –5°C and 19%
   (3) 2°C and 33%
   (4) –2°C and 33%   4 _____

5. A student used a sling psychrometer to measure the humidity of the air. If the relative humidity was 65% and the dry-bulb temperature was 10°C, what was the wet-bulb temperature?

   (1) 5°C
   (2) 7°C
   (3) 3°C
   (4) 10°C   5 _____

Base your answers to question 6 on the diagram, which shows a hygrometer located on a wall in a classroom. The hygrometer's temperature readings are used by the students to determine the relative humidity of the air in the classroom.

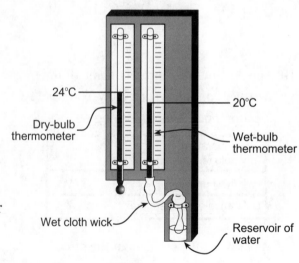

6. *a)* Based on the temperature readings shown in this diagram, determine the relative humidity of the air in the classroom. _____%

   b) Besides relative humidity, identify another weather variable of the air in the classroom that may be determined by using both temperature readings on the hygrometer.

   _____

   c) How does the wet cloth wick affect the wet-bulb thermometer?

   _____

7. What is the relative humidity when the air temperature is 28°C and the wet bulb temperature is 20°C?

   (1) 4%   (3) 19%
   (2) 16%   (4) 47%   7 _____

8. A sling psychrometer shows a dry-bulb reading of 14°C and a wet-bulb reading of 9°C. What are the dewpoint and the relative humidity?

   (1) –10°C and 16%
   (2) –10°C and 50%
   (3) 4°C and 16%
   (4) 4°C and 50%   8 _____

9. What is the difference between the dry-bulb temperature and the wet-bulb temperature when the relative humidity is 28% and the dry-bulb temperature is 0°C?

   (1) 11°C   (3) 28°C
   (2) 2°C   (4) 4°C   9 _____

10. The data below represent some of the weather conditions at a New York State location on a winter morning.

Air temperature (dry-bulb temperature)	0°C
Relative humidity	81%
Difference of temperature	?

   What was the dewpoint at this time?

   (1) 1°C   (3) –3°C
   (2) 2°C   (4) –5°C   10 _____

Base your answers to question 11 on the graph below, which shows the changes in relative humidity and air temperature during a spring day in Washington, D.C.

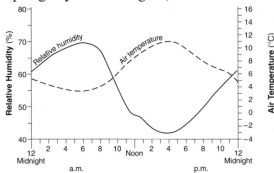

11. *a)* Which statement best describes the relationship between relative humidity and air temperature as shown by the graph?

   (1) Relative humidity decreases as air temperature decreases.
   (2) Relative humidity decreases as air temperature increases.
   (3) Relative humidity increases as air temperature increases.
   (4) Relative humidity remains the same as air temperature decreases.   a _____

   b) What were the relative humidity and air temperature at noon on this day?

   (1) 47% and 32°F
   (2) 65% and 32°F
   (3) 47% and 48°F
   (4) 65% and 48°F   b _____

   c) At which time would the rate of evaporation most likely be greatest?

   (1) 11 p.m.   (3) 10 a.m.
   (2) 6 a.m.   (4) 4 p.m.   c _____

   d) What is causing the air temperature to increase and the RH to decrease from 7 a.m. to 12 p.m.?

_____

12. Which station model represents a location that has the greatest chance of precipitation? (see page 158)

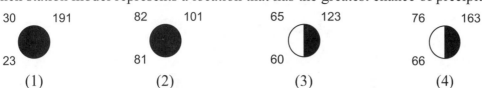

| (1) | (2) | (3) | (4) |

12 _____

13. What is the relative humidity within fog? _____

14. When you see a cloud are you seeing water vapor, or water droplets? _____

Base your answers to question 15 on the weather information below.

A student using a sling psychrometer obtained a dry-bulb reading of 20°C and a wet-bulb reading of 16°C for a parcel of air outside the classroom.

15. *a)* State the relative humidity. _____

b) State the change in relative humidity as the air temperature and the dewpoint temperature get closer to the same value. _____

c) Why does the wet bulb become colder than the dry bulb when the sling psychrometer is swung through unsaturated air?

_____

Base your answers to question 16 on the accompanying diagram, which shows the temperature change when a parcel of air warms, rises, and expands to form a cloud.

16. *a)* State the phase change that occurred to produced the cloud.

_____

b) State the relative humidity of the air at location *A*.

_____

c) What would cause the air parcel to expand as it rises?

_____

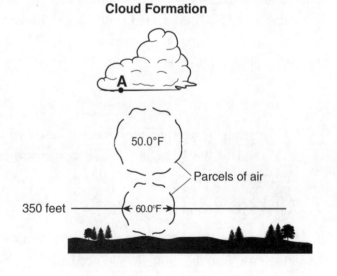

Cloud Formation

A

50.0°F

Parcels of air

350 feet ←60.0°F→

17. Explain why the relative humidity is usually different on the windward side of the mountain compared to the leeward side of a mountain.

_____

1. 3   The difference between the dry-bulb temperature and the wet-bulb temperature is 3°C. Locate this temperature difference at the top of the Relative Humidity chart. Move down in this column until you reach the Dry-Bulb Temperature row of 16°C. At this intersection point, the relative humidity is given as 71%.

2. 4   The difference between the dry-bulb temperature and the wet-bulb temperature is 6°C. Go to this temperature difference at the top of the Relative Humidity chart. Move down in this column until you reach the Dry-Bulb Temperature (which is the air temperature) row of 30°C. At this intersection point, the relative humidity is given as 61%. For odd number dry-bulb readings, the relative humidity answer will be found between two Dry-Bulb temperatures, in this case between 30°C and 28°C.

3. 3   The difference between the dry-bulb and wet-bulb temperature is 4°. Locate this temperature at the top of the chart and move down until the intersection of 2°C. The RH is given as 36%.

4. 3   The difference between the dry-bulb temperature and the wet-bulb temperature is 8°C. Move down from the 8°C Temperature Difference column to the Dry-Bulb Temperature row of 18°C for both charts. The dewpoint temperature is 2°C and the RH is 33%.

5. 2   Go down the Dry-bulb Temperature column until you reach 10°C. Move across this row to the RH value of 65%. It is located in the 3°C Difference Between Wet-Bulb and Dry-Bulb Temperatures column. If the dry-bulb temperature is 10°C and the difference is 3°C, the wet-bulb temperature, being colder, must be 7°C.

6. *a)*   Answer: 69%

Explanation: The difference between these two temperature readings is 4°C. Use this column found at the top of the RH chart. Move downward in this column to the intersection of the Dry-Bulb Temperature row of 24°C. At this intersection point the answer is given as 69%.

b)   Answer: Dewpoint temperature

Explanation: In order to get the dewpoint temperature and relative humidity, a wet-bulb thermometer and a dry-bulb thermometer are needed.

c)   Answer: Lowers its temperature

Explanation: As water evaporates from the wet cloth wick, it removes heat from the wet-bulb thermometer, lowering its temperature.

Temperature

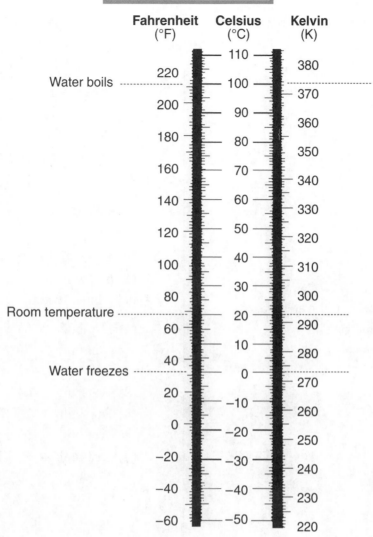

Fahrenheit (°F) Celsius (°C) Kelvin (K)

Water boils — 220 / 100 / 370

Room temperature — 60 / 20 / 290

Water freezes — 40 / 0 / 270

Overview:

We use the Fahrenheit scale, while all other countries use the Celsius scale. The Kelvin scale is mostly used by research scientists. The Fahrenheit and Celsius readings have units of °F and °C respectively. Readings in the Kelvin scale are assigned the letter K. All students should know the freezing point of water and boiling point of water on the Fahrenheit and Celsius scales. These readings are noted on this chart. Take time to learn them.

The Temperature Scales:

The calibration of each scale must be known. Each line on the Fahrenheit scale equals two degrees, while each line on the Celsius and Kelvin scale represents one degree. Knowing this, what is the boiling point of water on each scale at sea level? If located correctly, water's boiling point is 212°F, 100°C and 373 K.

Additional Information:

- Temperature is measured in degrees, and heat is measured in joules.

- As altitude increases, the boiling point temperature of water decreases because of a decrease in atmospheric pressure.

— Set 1 —

1. The highest air temperature ever recorded in Albany, New York, was 104°F, which occurred on July 4, 1911. This temperature is equal to

    (1) 35°C
    (2) 40°C
    (3) 45°C
    (4) 50°C   1 _____

2. At what temperature would water freeze on the Fahrenheit and Celsius temperature scales respectively?

    (1) 30°F and 0°C
    (2) 32°F and 0°C
    (3) 0°F and 32°C
    (4) 33°F and 1°C   2 _____

3. The human body temperature is 98.6°F. Give the human body temperature in the Celsius and Kelvin temperature scales.

    _____°C   _____K

4. At what temperature are the Celsius and Fahrenheit scale readings the same?

    _____

— Set 2 —

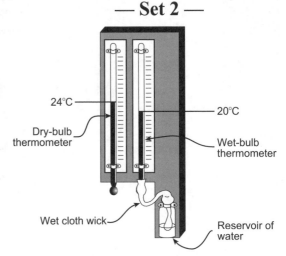

5. What would the dry-bulb and wet-bulb temperatures be when converted to Fahrenheit readings?

    (1) Dry-bulb — 68°F
        Wet-bulb — 76°F
    (2) Dry-bulb — 73°F
        Wet-bulb — 64°F
    (3) Dry-bulb — 76°F
        Wet-bulb — 68°F
    (4) Dry-bulb — 296°F
        Wet-bulb — 292°F   5 _____

6. Which scale has 100 divisions from the temperature when water freezes to the temperature when water boils?

    _____

7. What is the room temperature in all 3 temperature scales?

    _____°F   _____°C   _____K

8. Give the temperature that ice melts in all 3 temperature scales?

    _____°F   _____°C   _____K

Temperature

Answers

Set 1

1. 2 Using the Temperature chart, locate 104°F. Move directly across and the corresponding Celsius temperature is 40°C.

2. 2 Using the given Temperature chart, locate "Water Freezes" and obtain the correct readings. Remember, that these two temperature answers would be the same if the question asked: "What would be the melting point of ice on the Fahrenheit and Celsius scales?"

3. 37°C and 310 K (± 1 degree)

4. Answer: -40° (± 1 degree)

   Explanation: Go to -40° on the Fahrenheit scale and it is equal to -40° on the Celsius scale.

Air Pressure

Overview:

We live at the bottom of an "ocean" of air. Rarely do we ever feel the pressure it exerts upon us, but air has weight, producing pressure. Air pressure changes with changes in elevation and at times we feel this as our ears "pop." Even without any change in elevation, air pressure changes slightly due to temperature and moisture changes. The instrument used to measure air pressure is the barometer. In the mercury barometer, the mercury (Hg – see bottom of the chart) is held up by the atmospheric pressure to a height of around 30 inches. This height, reflecting the atmospheric pressure, fluctuates with changing air temperature and moisture content. Typically, if the air pressure is steadily dropping, it signifies that a low-pressure system (L) full of moisture is moving in. A steadily rising barometer usually indicates a high-pressure system (H) is approaching, bringing in cooler, drier air. Very low-pressure readings are associated with hurricanes.

The Chart:

Pressure readings are recorded in either millibars (mb) or inches (in) of Hg. Each line on the millibar scale represents 1 mb, while each line on the inches scale represents 0.01 in. The "One atmosphere" reading is shown on this chart as 1013.2 mb, which closely equals 29.92 in. What does 996.0 mb equal in inches? If done correctly the answer is 29.41 in.

The millibar pressure readings are abbreviated, when placed on a weather station model, by dropping the decimal and either the 9 or the 10 number found at the front of the pressure reading. Using this rule, 1013.6 mb becomes 136 and a reading of 998.4 mb becomes 984. To change a station model number back to the actual pressure reading, use the 500 rule. If a number is lower than 500, add a 10 and replace the decimal. If a pressure reading is higher than 500, add a 9 and the decimal. For example, a station model barometric reading of 348 is less than 500, adding a 10 and replacing the decimal converts it to 1034.8 mb. For a station model reading of 978, which is higher than 500, adding a 9 and replacing the decimal converts it to 997.8 mb.

Additional Information:

- Warm air rises, producing a lower barometric pressure reading.
- Cold air sinks, producing a higher barometric pressure reading.
- Isobars connect areas of equal air pressure values.
- A strong air pressure gradient will produce windy conditions.
- Wind blows from H pressure regions to L pressure regions.
- When water vapor molecules move in, they displace air molecules causing the barometer pressure to fall. This occurs because the composition of water vapor is 66% hydrogen – the lightest of all elements.

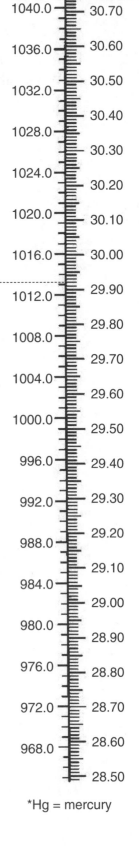

millibars (mb)	inches (in of Hg*)
1040.0	30.70
1036.0	30.60
	30.50
1032.0	30.40
1028.0	30.30
1024.0	30.20
1020.0	30.10
1016.0	30.00
One atmosphere	
1012.0	29.90
	29.80
1008.0	29.70
1004.0	29.60
1000.0	29.50
996.0	29.40
992.0	29.30
	29.20
988.0	29.10
984.0	29.00
980.0	28.90
976.0	28.80
972.0	28.70
	28.60
968.0	28.50

*Hg = mercury

Diagrams:

1. **Mercury Barometer** – A barometer measures the atmospheric air pressure in millibars (mb) or inches (in) of mercury (Hg). The mercury in a barometer tube is held up by air pressure. As the air pressure changes, the height of the mercury fluctuates. Moist, warmer air produces lower barometric readings, while cold, dry air produces higher barometric readings. An aneroid barometer (see question 18, page 156) works on a mechanical mechanism that moves a dial as pressure changes. Barometric readings are used in weather forecasting and can also be used to determine altitude above sea level.

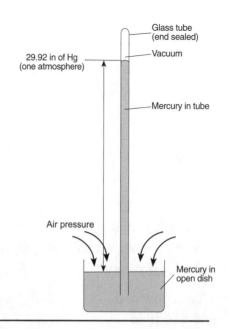

2. **Air Pressure Recording** – Air pressure is affected by moisture and temperature. A barometer recorded the air pressure over 3 days when a strong L pressure system moved through. As the L pressure system entered, having much moisture, the pressure dropped to 960.0 mb and the station recorded rain. Cold and drier air followed, producing higher pressure readings. Hurricanes produce very low pressure reading.

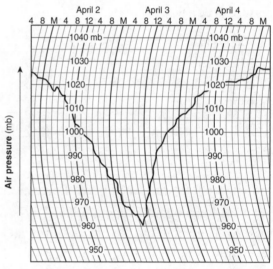

3. **Land and Sea Breezes** –
Sea Breeze – During the day, the hotter, less dense air over the land rises, producing L pressure and the colder, denser, sinking air over the water produces H pressure. This difference in pressure sets up a sea breeze.

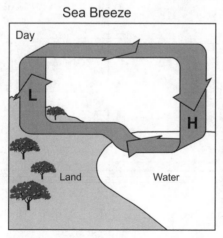

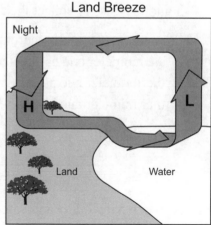

Land Breeze – At night, the land temperature is lower than the warmer water. Now the H pressure is over the cooler land, and the L pressure, caused by the rising air, is over the water. This air circulation pattern is a land breeze. Both of these breezes are local air movements.

1. A pressure of 1036.0 mb when converted to inches would equal

   (1) 30.50 in   (3) 30.60 in
   (2) 30.59 in   (4) 30.55 in   1 _____

2. The diagram below represents an aneroid barometer that shows the air pressure, in inches of mercury.

   When converted to millibars, this air pressure is equal to

   (1) 1009.0 mb
   (2) 1012.5 mb
   (3) 1015.5 mb
   (4) 1029.9 mb   2 _____

3. Students wish to study the effect of elevation above sea level on air temperature and air pressure. They plan to hike in the Adirondack Mountains from Heart Lake, elevation 2,179 feet, to the peak of Mt. Marcy, elevation 5,344 feet. Which instruments should they use to collect their data?

   (1) anemometer and psychrometer
   (2) anemometer and barometer
   (3) thermometer and psychrometer
   (4) thermometer and barometer   3 _____

4. A pressure reading of 29.50 inches when converted to millibars would equal

   (1) 996.3 mb   (3) 999.9 mb
   (2) 996.6 mb   (4) 999.0 mb   4 _____

5. A dedicated earth science student using an aneroid barometer measures the air pressure at the base of Slide Mountain as 1016.0 mb. At the same time another dedicated earth science student measures the pressure at the top of the mountain. Which pressure reading could be correct at the top of Slide Mountain?

   (1) 1016.0 mb   (3) 1025.7 mb
   (2) 988.5 mb   (4) 1040.0 mb   5 _____

6. Weather-station measurements indicate that the dewpoint temperature and air temperature are getting farther apart and that air pressure is rising. Which type of weather is most likely arriving at the station?

   (1) a snowstorm
   (2) a warm front
   (3) cool, dry air
   (4) maritime tropical air   6 _____

7. An air pressure reading of 987.5 mb would be placed on a station model as

   (1) 987   (3) 875
   (2) 87.5   (4) 750   7 _____

8. Explain what would happen to the barometric reading if more moisture is entering into the surrounding air.

_____

_____

Note: For some questions, you will need to use the Station Model Explanation chart on page 158.

9. Which set of air pressure readings are the same?

(1) 1004.0 mb and 29.90 inches
(2) 992.0 mb and 29.29 inches
(3) 1016.0 mb and 30.10 inches
(4) 1000.5 mb and 29.95 inches   9 _____

Note: Question 10 has only three choices.

10. At 9:00 a.m. the atmospheric pressure was 1002.6 mb and air temperature was 63°F. Four hours later the temperature rose to 93°F. If there was no change in elevation and in the atmospheric moisture level, the expect barometer reading should have

(1) increased
(2) decreased
(3) remained the same   10 _____

11. The Moon lacks an atmosphere. Of the following statements, which is correct?

(1) Air pressure would still exist since the Moon has gravity.
(2) Air pressure would still exist since the Earth's atmosphere extends to the Moon.
(3) Air pressure would not exist on the Moon.
(4) Air pressure on the Moon equals the Earth's air pressure.   11 _____

12. A station model air pressure reading of 058 would convert to

(1) 905.8 mb   (3) 1005.8 mb
(2) 105.8 mb   (4) 10005.8 mb   12 _____

13. A station model barometric pressure reading of 996 would convert to

(1) 996.0 mb   (3) 1099.6 mb
(2) 999.6 mb   (4) 9990.6 mb   13 _____

14. An air pressure of 1003.6 mb would be placed on a station model as

(1) 1003   (3) 03.6
(2) 100   (4) 036   14 _____

15. A station model is shown to the right. What is the air pressure at this location?

(1) 902.9 mb   (3) 1029.0 mb
(2) 1002.9 mb   (4) 9029.0 mb   15 _____

16. Which weather-station model shows an air pressure of 993.4 millibars.

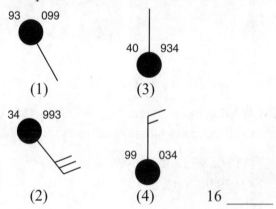

17. Close spacing of isobars on a weather map is a good indicator of

(1) low visibility
(2) low dewpoint temperatures
(3) high air temperatures
(4) high wind velocity   17 _____

18. A weather instrument is shown below.

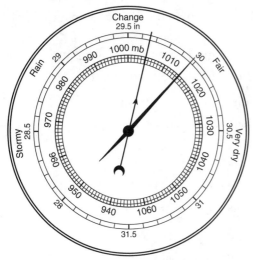

Which weather variable is measured by this instrument?

(1) wind speed   (3) cloud cover
(2) precipitation   (4) air pressure   18 _____

19. The cross section below shows a sea breeze blowing from the ocean toward the land. The air pressure at the land surface is 1013 millibars.

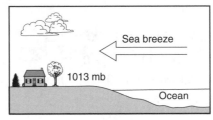

(Not drawn to scale)

The air pressure at the ocean surface a few miles from the shore is most likely

(1) 994 mb   (3) 1013 mb
(2) 1005 mb   (4) 1017 mb   19 _____

20. Which cross section below best shows the locations of high air pressure and low air pressure near a beach on a hot, sunny, summer afternoon?

	Key
H	High air pressure
L	Low air pressure

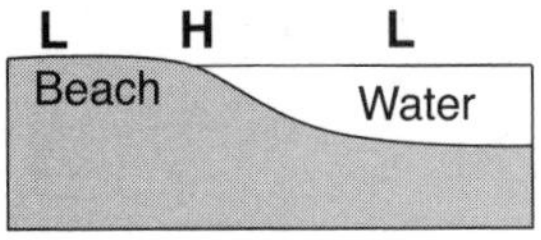

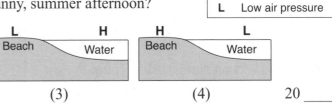

(1) 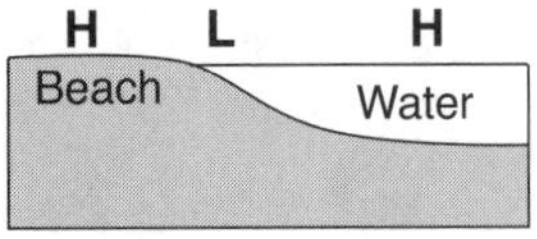   (2)   (3)   (4)   20 _____

21. What is the average air pressure exerted by Earth's atmosphere at sea level, expressed in millibars and inches of mercury?

(1) 1013.25 mb and 29.92 in of Hg   (3) 1012.65 mb and 29.91 in of Hg
(2) 29.92 mb and 1013.25 in of Hg   (4) 29.91 mb and 1012.65 in of Hg   21_____

22. Give a statement on the movement of air between a H pressure and a L pressure system.

_____

23. When dry, cold Canadian air displaces very humid air, what change occurs to the air pressure?

_____

1. 2   Using the Air Pressure chart, locate the 1036.0 mb line. Read across to the inches scale, one finds the value of 30.59 in. Remember, that each line on the inches scale represents 0.01 in.

2. 2   The barometric reading from the aneroid barometer is 29.90 in. Find this number on the Air Pressure chart. Read directly across one locates 1012.5 mb.

3. 4   A thermometer measures air temperature. A barometer measures air pressure and is sensitive to elevation changes. An anemometer measures wind speed.

4. 4   On the Air Pressure chart, locate the 29.50 inch line. Read directly across to the millibars scale to 999.0 mb.

5. 2   As altitude increases, the air pressure decreases. As the dedicated earth science student ascends the mountain, the pressure will decrease. This occurs because there is less air at the top of a mountain.

6. 3   When the air and dewpoint temperatures are getting farther apart, the relative humidity will decrease, indicating dry air is moving in. Cold air, being dense, will produce rising air pressure. Remember, the highest air pressure (at sea level) is when the air is cold and dry.

7. 3   From a barometer reading, the 9 or 10 is dropped as well as the decimal to make a station model reading. Thus, 987.5 mb becomes 875.

8. Decreases *or* drops

   Explanation: When moisture enters the air, the $H_2O_{(g)}$ molecules displace air molecules causing the pressure to decrease. These water vapor molecules have less weight than the same volume of air molecules.

Remember:

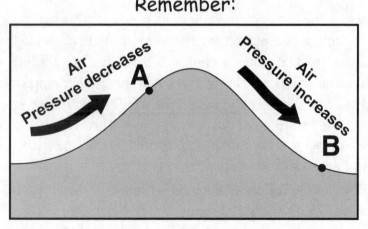

Key to Weather Map Symbols

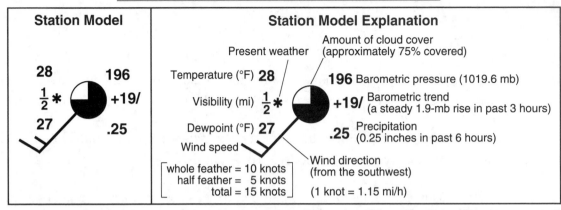

Station Model

Station Model Explanation

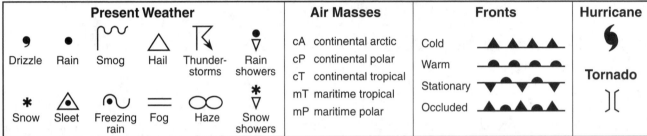

Present Weather						Air Masses		Fronts		Hurricane

Present Weather

Drizzle   Rain   Smog   Hail   Thunder-storms   Rain showers

Snow   Sleet   Freezing rain   Fog   Haze   Snow showers

Air Masses

cA   continental arctic
cP   continental polar
cT   continental tropical
mT   maritime tropical
mP   maritime polar

Fronts

Cold
Warm
Stationary
Occluded

Hurricane

Tornado

Overview:

If you don't like the weather, just wait an hour; this is so true for NYS weather! The different weather variables, such as temperature, cloud cover, air pressure, wind speed, etc., that change constantly are recorded on a station model. These station models, sent from hundreds of cities throughout United States to the National Weather Service, are recorded on weather maps and are used to assist in weather forecasting. Station models have no words or units, just numbers and symbols. These numbers and symbols must be placed in their assigned location as shown by the above Station Model Explanation. Once the station models are placed on weather maps, air masses and their fronts become evident. An air mass is a large body of air that has similar properties throughout. An air mass moves due to atmospheric pressure and is affected by the rotation of the Earth. There are five different air masses classified by their relative temperature and moisture condition. At the boundaries of these air masses, four fronts are possible. Here unstable weather is usually found as one air mass clashes with another. So whether you're ready, or whether you're not, let's get going on building weather station models!

The Charts:

Station Model Explanation – The center circle represents the city that is reporting the weather. This circle is shaded to show the percentage of cloud cover. The wind speed is recorded in knots, and the direction is where the wind is coming from. To show a north wind, a shaft (straight line) is drawn from the north into the circle. The wind speed is indicated by feathers. A whole feather represents 10 knots, and a half of a feather represents 5 knots. Make sure you round off the wind speed to the nearest 5 knots. For example, a wind speed of 27 knots would be rounded down to 25 knots. This would be shown by two and one-half feathers. A wind speed of 38 knots, being rounded up to 40 knots, would have 4 full feathers. The air and the dewpoint temperatures must be in Fahrenheit, but the units of °F are not placed on the station model. For the present weather, select the correct symbol that is found in the Present Weather box. Visibility is how far the observer can see and is measured in miles. The barometer reading is always abbreviated on a station model. This is done by dropping the decimal and either the 10 or the 9 found at the front of the barometric reading. For example, a barometric reading of 1036.3 mb becomes 363 and a barometer reading of 997.3 mb becomes 973 on a station model. The barometric trend is what has happened to the pressure over the past 3 hours. For this reading, the decimal is dropped, and a line is used to show the pressure trend. For example, if the air pressure has dropped 2.8 mb in the last 3 hours, it is recorded on the station model as –28 \ . The precipitation amount is for the past 6 hours. Remember, on a station model you must place the number or symbol in its correct designated position.

Station Model – This is how a completed station model would appear, ready to be inserted on a weather map. Notice, no units or words are placed on a station model.

Present Weather – One of these symbols would be placed on the station model if this current weather condition is occurring. Sleet and freezing rain are different. Sleet is ice pellets created as rain freezes as it falls. Freezing rain is rain that freezes into glaze upon contact with the frozen ground.

Air Masses – These symbols identify the properties within an air mass. Each air mass is assigned two letters. One letter indicates the relative moisture content, and the other letter indicates the relative temperature of the air mass. The moisture letters are either m or c. The m stands for maritime, representing much moisture, having its source area over water. The c stands for continental, which indicates that the air mass is relatively dry, having its source area over land. For temperature, T stands for tropical (hot), P stands for polar (cold), and A stands for arctic (very cold). From these letters, 5 sets of combinations are possible based on temperature and moisture properties of the air mass. These combinations are shown in the Air Masses section.

Front Symbols – A front is the boundary area between two different air masses. At the front, the moist, warmer air is being forced to rise over the drier, denser, colder air. This rising air will cool and reach its dewpoint temperature. Now water vapor condenses to form clouds, and usually some type of precipitation occurs. As shown, there are four different front symbols. If you see the symbol for a cold front on a weather map, the cold air is behind the front. The same is true for a warm front; the warm air is behind the front. A stationary front is not moving and may bring a long period of precipitation. The last front is called an occluded front; it too, is usually associated with a long period of precipitation.

Hurricane and Tornado Symbols – The hurricane symbol will be used when winds of a tropical storm become higher than 75 mph, making it a hurricane. If you have been vacationing near the coastal waters, you might have noticed this symbol on hurricane evacuation signs. Remember that a hurricane is a large, strong L pressure system that may last for weeks. It is assigned the air mass letters of mT. A tornado is an extreme low-pressure, high-speed funnel cloud that spins counterclockwise. A tornado is short lived and affects a narrow geographic area, doing great damage.

Additional Information:

- Because a hurricane is a low-pressure system, it spins counterclockwise around its eye. The energy of a hurricane is obtained from the warm ocean water. When it passes over land, it loses its energy source and becomes weaker, resulting in diminishing wind speed.

- Monsoons are caused by unequal heating rates of land and water. This, along with a seasonal shift in wind direction, produces excessive rainfall in many parts of the world, most notably India.

Diagrams:

1. **Weather Instruments** – An anemometer (1) is an instrument that measures the wind speed. A wind vane (2) is an instrument that shows the direction of the wind.

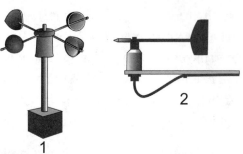

2. **Weather Station Models** – Station model *A* shows 7 weather variables: wind speed of 25 knots, wind direction of NW, cloud cover of 100%, air temperature of 72°F, dewpoint temperature of 72°F, a barometric air pressure of 999.6 mb, and the presence of thunderstorms. Station *A* has a relative humidity of 100% because the dewpoint and air temperatures are the same.

Station *B* is reporting a barometric air pressure of 1002.2 mb, a 20-knot wind from the NW, 25% cloud cover, an air temperature of 64°F, and a dewpoint of 58°F. The visibility is 8 miles. Remember, no units (mb, °F, mi, etc.), or words are placed on station models.

Station A

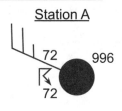

Station B

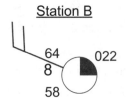

3. **Surface Wind Circulation in Pressure Systems** – The surface air movement in a low (L) pressure system circulates counterclockwise and inward toward the center. This air then rises, forming clouds.

The surface air movement in a high (H) pressure system circulates clockwise and outward from the center of the system. The air at the center is sinking downward toward the surface. Sinking air heats up, reducing the chance of cloud formation.

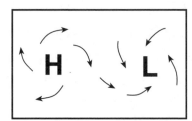

4. **Isobars** – Isobars are isolines that connect areas of equal air pressure measured in millibars (mb). From isobars, H and L pressure centers can be identified. L pressure regions are associated with "unsettling weather", normally having some type of precipitation. H pressure regions are associated with clear skies. The closer the isobars are to each other, the greater the pressure gradient will be, causing high wind speeds. Location *B* is experiencing higher wind speeds than location *A*.

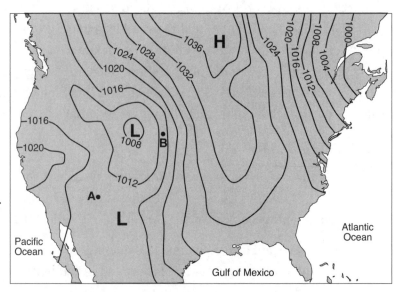

Key to Weather Map Symbols

5. **Fronts** – This diagram shows 3 different fronts, a cold front extending down from *B* to *C*, a warm front from *B* to *D*, and an occluded front from *B* to *A*. Behind the warm front is situated the warm, moist air (mT); behind the cold front is situated the cold, dry air (cP).

Along all fronts the air is unstable, usually rising and producing some type of precipitation. These fronts will advance NE making their presence known in NYS shortly.

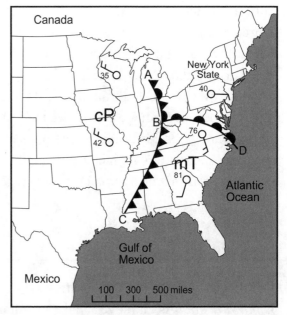

6. **Cross Section of a Cold Front** – The air within the cP (continental polar) air mass is cold and dry, making it denser than the warmer, moist mT (maritime tropical) air mass. When the cP air mass advances, the less dense, warmer mT air is displaced upward. When air rises it expands, cools, and hits its dewpoint temperature. Now, water vapor condenses, forming clouds, and eventually precipitation occurs.

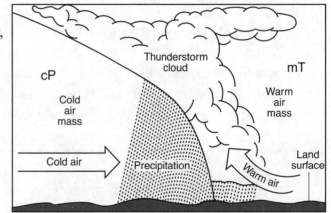

7. **Cross Section of Frontal Boundaries** – The cold dry (cP) air mass (*A*) is displacing the warmer moist (mT) air mass (*B*) which is displacing another cP air mass (*C*). At the frontal surfaces, *X* and *Y*, warmer air is being forced upward over the colder air. As this air rises, clouds form and precipitation occurs. In time, these air masses and their associated fronts will advance, influencing weather conditions as they move along.

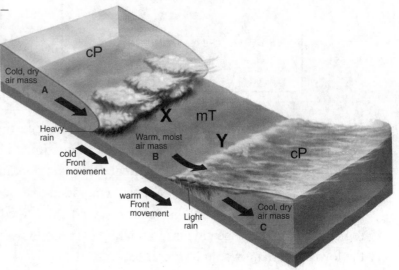

8. **Source Regions of Air Masses** – The geographic source region of an air mass is where it originated. This cP (continental polar) air mass was formed in the dry, cold interior of Canada. The shown mT (maritime tropical) air mass picked up its warm, moist properties from the Gulf of Mexico. When their advancing fronts converge, major weather changes occur that might include violent thunderstorms in which tornadoes could develop.

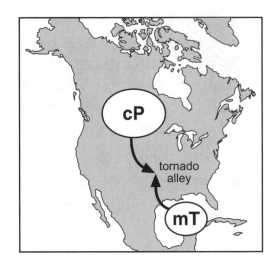

9. **Windward and Leeward Conditions** –

Windward side: As air moves up the mountain, it expands, cools, and hits its dewpoint temperature. Now clouds form producing precipitation.

Leeward side: The now drier air descends and starts to be compressed. Compressing air causes the temperature to increase. These conditions produce little chance of rain, producing a much drier and hotter climate compared to the windward side of the mountain.

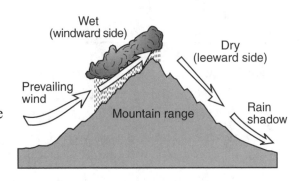

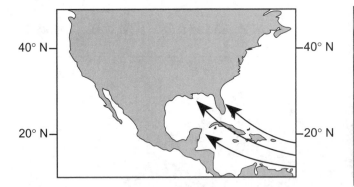

10. **Hurricane** – A hurricane is a very strong, large, long-lasting low (L) pressure system. This system acquires its energy from warm ocean water. Many times a hurricane's path will bring it into the Gulf of Mexico causing much wind, tidal surge and flooding damage to the affected coastal states. As it loses its energy over land, it gets downgraded to a tropical depression. Much rain continues, causing streams to reach flood-stage levels, and extensive flooding occurs, especially within floodplain regions. Notice how the clouds are rotating counterclockwise around the eye of the hurricane where the lowest pressure would be recorded.

Key to Weather Map Symbols

1. Which weather-station model shows an air pressure of 993.4 millibars?

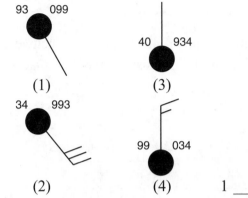

1 _____

2. Which station model correctly shows the weather conditions of a thunderstorm with heavy rain?

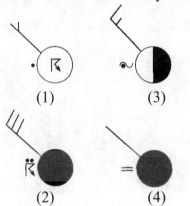

2 _____

3. The properties of an air mass are mostly determined by the

(1) rate of Earth's rotation
(2) direction of Earth's surface winds
(3) source region where the air mass formed
(4) path the air mass follows along a land surface

3 _____

4. A temperature reading of 32°F would be shown on a station model as

(1) 0°C   (3) 32°F
(2) 0   (4) 32

4 _____

5. Which air mass is normally associated with the formation of hurricanes?

(1) continental tropical
(2) maritime tropical
(3) continental polar
(4) maritime polar

5 _____

6. Which type of air mass would most likely have low humidity and high air temperature?

(1) cT   (3) mT
(2) cP   (4) mP

6 _____

7. A weather station model for a location in New York State is shown. The air mass over this location is best described as

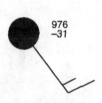

(1) cold with low humidity and high air pressure
(2) cold with high humidity and low air pressure
(3) warm with high humidity and low air pressure
(4) warm with low humidity and high air pressure

7 _____

8. The station model shows the weather conditions at Massena, New York, at 9 a.m. on a particular day in June. What was the barometric pressure at Massena 3 hours earlier on that day?

(1) 997.1 mb   (3) 1003.3 mb
(2) 999.7 mb   (4) 1009.1 mb

8 _____

9. In New York State, dry, cool air masses (cP) often interact with moist, warm air masses (mT). Which statement correctly matches each air mass with its usual geographic source region?

(1) cP is from the North Atlantic Ocean and mT is from the deserts of the southwestern United States.
(2) cP is from northern Canada and mT is from the deserts of the southwestern United States.
(3) cP is from northern Canada and mT is from the Gulf of Mexico.
(4) cP is from the North Atlantic Ocean and mT is from the Gulf of Mexico.

9 _____

10. The accompanying map shows the boundary between two air masses. The arrows show the direction in which the boundary is moving.

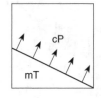

Which weather map uses the correct weather front symbol to illustrate this information?

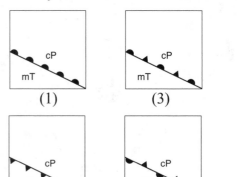

(1)   (3)

(2)   (4)   10 _____

Base your answers to question 11 on the accompanying map.

11. *a)* What is the total number of different kinds of weather fronts shown on this weather map?

(1) 1   (3) 3
(2) 2   (4) 4   a_____

b) The air mass influencing the weather of Nebraska most likely originated in

(1) the northern Pacific Ocean
(2) the northern Atlantic Ocean
(3) central Canada
(4) central Mexico   b_____

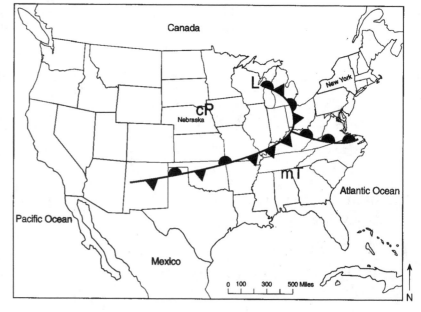

12. Hurricane season in the North Atlantic Ocean officially begins in June and ends in November. Which ocean surface conditions are responsible for the development of hurricanes?
(1) warm water temperatures and low evaporation rates
(2) warm water temperatures and high evaporation rates
(3) cool water temperatures and low evaporation rates
(4) cool water temperatures and high evaporation rates

12 _____

13. Which map best shows the general surface wind pattern around the high-pressure system?

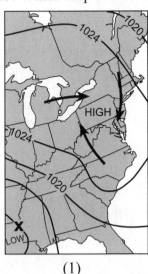 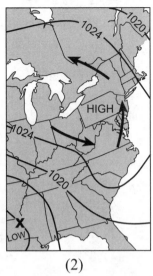 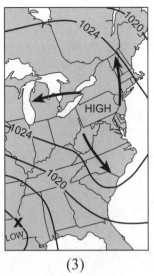 

(1)   (2)   (3)   (4)

13_____

14. A weather station at a lighthouse records foggy conditions, temperature of 36°F, winds from the SE at 15 knots, and an air pressure of 1016.4 mb. Using the proper weather map symbols, place the following information in the correct positions on the weather station model.

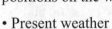

• Present weather   • Air pressure   • Dewpoint
• Air temperature   • Wind direction   • Wind speed

Base your answers to question 15 on the accompanying weather station model.

15. *a)* State the condition represented by the symbol for "present weather." _____

b) State the full barometric pressure. _____

c) State the visibility. _____

d) State the cloud coverage. _____

e) State the wind speed and its direction. _____

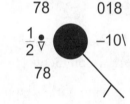

78   018

$\frac{1}{2}$ • ∇   −10\

78

Base your answers to question 16 on the accompanying weather map. The shaded portion represents an area of precipitation.

16. *a)* What type of front extends eastward from the low-pressure center?_____

b) What is the wind speed and direction for Albany? _____

c) What is the relative humidity at Albany? _____%

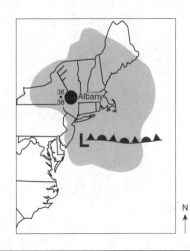

Base your answers to question 17 on the weather map of North America below. The map shows the location of a front and the air mass influencing its movement.

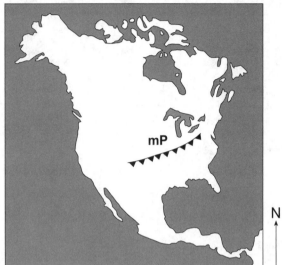

17. *a*) Which region is the probable source of the air mass labeled mP on the map?

(1) central Canada
(2) southwestern United States
(3) Gulf of Mexico
(4) Gulf of Alaska   a _____

b) Which type of front and frontal movement is shown on the weather map?

(1) cold front moving northwestward
(2) cold front moving southeastward
(3) warm front moving northwestward
(4) warm front moving southeastward

b _____

c) The mP air mass is identified on the basis of its temperature and

(1) wind direction
(2) cloud cover
(3) moisture content
(4) wind speed   c _____

18. An air mass classified as mT usually forms over which type of Earth surface?

(1) warm land   (3) cool land
(2) warm ocean   (4) cool ocean   18 _____

19. What is the visibility, in miles, shown on the station model?

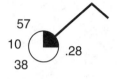

(1) 10   (3) 38
(2) 28   (4) 57   19 _____

20. Various weather conditions at LaGuardia Airport in New York City are shown on the station model.

What were the barometric pressure and weather conditions at the airport at the time of the observation?

(1) 914.6 mb of pressure and smog
(2) 914.6 mb of pressure and a clear sky
(3) 1014.6 mb of pressure and smog
(4) 1014.6 mb of pressure and a clear sky   20 _____

21. Which weather map symbol is associated with extremely low air pressure?

(1)   (2)   (3)   (4)   21 _____

22. An air mass entering Alaska from the northern Pacific Ocean would most likely be labeled on a weather map as

(1) cP   (3) mP
(2) cT   (4) mT   22 _____

23. Which type of air mass most likely has high humidity and high temperature?

(1) cP   (2) cT   (3) mT   (4) mP   23_____

24. Which station model represents a city with a relative humidity of 100%?

(1)   (2)   (3)   (4)   24_____

25. The accompanying map shows partial weather conditions for weather stations *A* and *B* at 4 p.m. A weather front is located between the two stations.

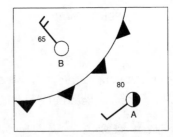

Which graph represents the temperature change that will most likely occur at station *A* as the front passes in the next three hours?

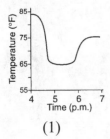 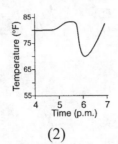 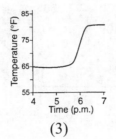 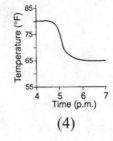

(1)   (2)   (3)   (4)   25_____

26. Which sequence of events affecting moist air within Earth's atmosphere causes cloud formation?

(1) rising expanding cooling condensation   (3) sinking expanding warming condensation
(2) rising contracting warming evaporation   (4) sinking contracting cooling evaporation   26_____

Base your answers to question 27 on the meteorological conditions shown in the table and partial station model, as reported by the weather bureau in the city of Oswego, New York. The diagram of the station model appears next to it.

Air temperature: 65°F
Wind direction: from the southeast
Windspeed: 20 knots
Barometeric pressure: 1017.5 mb
Dewpoint: 53°F

27. *a)* Using the meteorological conditions given, complete the station model above.

b) State the sky conditions or amount of cloud cover over Oswego as shown by the station model. _____

c) How would the pressure reading be shown on a station model?_____

28. On the weather map station model below, using the proper format, record the seven weather conditions shown below.

Wind: from the northwest
Wind speed: 35 knots
Barometric pressure: 1022.0 mb
Cloud cover: 50%
Visibility: 5 mi
Precipitation (in the past 6 hours): .45 in
Present weather: Fog

29. What is meant by source region?_____

Base your answers to question 30 on the information on the four station models shown below. The weather data were collected at Niagara Falls, Syracuse, Utica, and New York City at the same time.

30. *a*) What is the air pressure in Syracuse? _____mb

b) Explain how the weather conditions shown on the station models suggest that Utica has the greatest chance of precipitation.

_____

c) Which city has the lowest air pressure? _____

d) What was the barometric pressure for Niagara Falls 3 hours ago? _____mb

e) New York City was experiencing a wind blowing from the south at 10 knots with hazy conditions limiting visibility to ³/4 of a mile. On the station model for New York City below, place in the proper location and format, the information below.

• wind direction   • wind speed   • present weather   • visibility

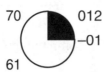

f) Give the full barometric pressure reading from the above station model. _____ mb

Key to Weather Map Symbols

31. Describe how the air's temperature and water vapor content at point X is different from the air's temperature and water vapor content at point Y.

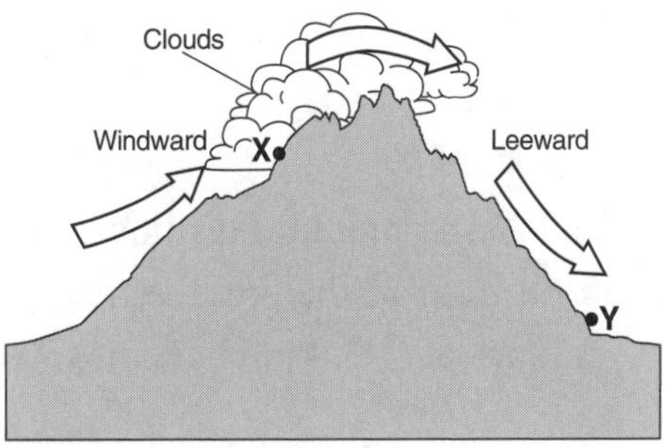

Air temperature:_____

Water vapor content: _____

Base your answers to question 32 on the cross section, which shows a typical cold front moving over New York State in early summer.

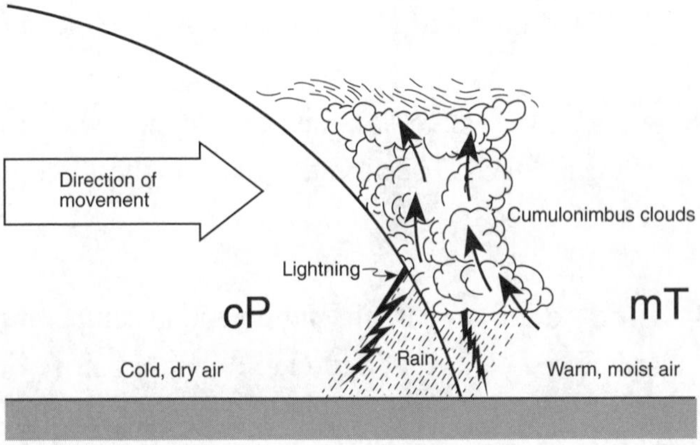

32. *a*) Explain why the warm, moist air is rising at the frontal boundary.

_____

b) State *one* process that causes clouds to form in this rising air. _____

c) Central Canada was the geographic source region for the cP air mass shown in the cross section. Identify the most likely geographic source region for the mT air mass shown in the cross section.

_____

d) When the cold front passes, what will occur to the barometric pressure?

_____

33. An Earth Science class is preparing a booklet for emergency preparedness. State two safety measures that should be taken to minimize danger from each of the following treats.

a) Hurricane 1 _____

    2 _____

b) Tornado 1_____

    2_____

34. Describe one ocean surface condition or atmospheric condition that makes the area over the Atlantic Ocean between 10° N latitude and 20° N latitude ideal for hurricanes to form.

_____

1. **3**   Barometric pressure readings are changed to a station model format by dropping the 10 or 9 found at the front of the pressure readings and dropping the decimal. Using this method, the barometric pressure of 993.4 mb becomes 934 on a station model.

2. **2**   Choice 2 shows the correct location for "present weather", a thunderstorm (see Station Model Explanation, page 158). We must assume that a double dot indicates heavy rain.

3. **3**   All air masses are assigned two letters, which identifies the moisture and temperature properties of the air mass. The air mass picks up these properties from where it was formed, which is its source region.

4. **4**   On a station model, air and dewpoint temperatures are recorded in Fahrenheit. No units or words are placed on a station model.

5. **2**   Hurricanes are very strong, large, warm, moisture-saturated L pressure systems. Their source areas are in warm tropical waters. They are classified as an mT air mass to indicate the high moisture content (m = maritime) and warm temperature (T = tropical).

6. **1**   A dry air mass would have low humidity. The origin of such an air mass would be over land and is assigned the letter c for continental – indicating a dry air mass. A warm air mass is assigned the letter T, which stands for tropical.

7. **3**   The temperature of 81°F makes the air mass warm. The station model pressure of 976 becomes 997.6 mb. Open to the Air Pressure chart (see page 152). This reading is near the bottom, making it a low pressure air mass. When the air and dewpoint temperatures are close to each other, the relative humidity is high.

8. **3**   The station model barometric pressure of 002 represents a pressure reading of 1000.2 mb. The barometric trend, which is measured over the past 3 hours, is -31\. This barometric reading had the decimal dropped, so the actual pressure trend is -3.1 mb. The negative sign shows that the pressure has fallen. So, 3 hours ago, the air pressure was 3.1 mb higher or 1000.2 + 3.1 = 1003.3 mb.

9. **3**   A cP air mass is a dry (c), cold (P) air mass that could have originated in central northern Canada. An mT air mass is a wet (m), warm (T) air mass that could have originated in the Gulf of Mexico.

10. **1**   The warm mT air mass is heading northeast, pushing out the cP air mass. Since the mT air mass is moving in, the warm front symbol is used. Remember that the warm air is behind the warm front.

11. *a)* 4 Locate the Fronts chart in the Weather section. All four of these fronts symbols are on the map. Next to the L is an occluded front. The warm front is heading toward the northeastern states. The cold front is dropping down, moving into Tennessee and Kentucky. The stationary front extends through Oklahoma and Texas.

   b) 3 A cP air mass sits over Nebraska. These letters signify that this air mass is dry (c = continental) and cold (P = polar). The source region for these conditions exists in central Canada.

12. 2 Hurricanes form over warm tropical waters where much evaporation will occur. When water evaporates, it absorbs much solar energy. This energy is later released by condensation and is a huge source of energy for hurricanes.

13. 4 The surface winds in a high pressure system rotate clockwise and outward.

14. Explanation: The present weather is fog, thus the relative humidity is 100%. When this occurs the dewpoint temperature and air temperature must be the same. The feathers for wind speed may be placed on either side of the shaft.

15. *a)* Answer: Rain showers

   Explanation: The present weather symbol is always placed between the visibility area and the circle. Match the present weather symbol in the question to those given in the Present Weather box.

   b) Answer: 1001.8 mb

   Explanation: When the air pressure on a station model is less than 500 add a 10 and the decimal to convert it back to the correct full barometric pressure; therefore, 018 = 1001.8 mb.

   c) Answer: $\frac{1}{2}$ mile

   Explanation: See Station Model for position of visibility.

   d) Answer: 100%

   Explanation: The circle is fully shaded in, which represents total cloud coverage of the sky.

   e) Answer: 5 knots from the SE

   Explanation: Half of a feather = 5 knots, the wind shaft is from the SE.

16. *a)* Answer: Occluded front

   Explanation: Match the given front to the ones shown in the Fronts section of the Weather Key area.

   b) Answer: 25 knots from the northwest

   Explanation: Each full feather is 10 knots, half of a feather is 5 knots. The wind direction is always where the wind is coming from. The wind shaft shows that the wind is from the NW.

   c) Answer: 100%

   Explanation: The station model for Albany shows that the air temperature and dewpoint temperature are equal. When this occurs, the relative humidity will be 100%, and the air is saturated with moisture.

Selected Properties of Earth's Atmosphere

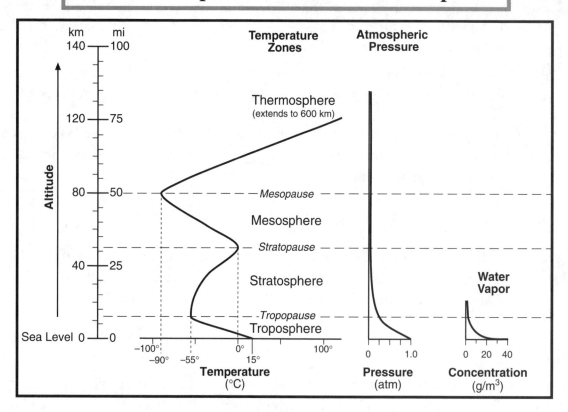

Overview:

Every mountain climber knows that as one ascends a mountain, atmospheric conditions quickly change. This chart covers three properties that change as altitude increases: temperature, atmospheric pressure, and water vapor. As altitude increases, there is less air, producing less air pressure. The amount of water vapor also decreases with altitude. Temperature is different; it reverses itself more than once. These temperature changes are the reason why the atmosphere is assigned different layers.

The Chart:

Altitude scale – On the left is the Altitude scale in kilometers (km) and miles (mi). Each interval (line) on the kilometer scale represents 10 km, and each line on the mile scale represents 5 mi. The Sea Level 0 line is the Earth's surface, being at the bottom of the atmosphere.

Temperature Zones graph – The Temperature Zones graph shows the four different layers into which the atmosphere is divided. The suffix "pause" indicates the transition of one layer to the next. The horizontal dashed lines represent the top of each layer and the beginning of the next atmospheric layer. The first layer, which we live in, is the troposphere. In this layer, as the altitude increases, the temperature decreases. At top of the troposphere (the tropopause), the temperature is shown to be –55°C. Passing through the tropopause, one enters the next layer, the stratosphere. The dash line shows that this occurs approximately at the 12 km or 7 mi level. In the stratosphere, the temperature reverses and warms up to 0°C at the stratopause. Once again, the temperature reverses and becomes colder in the mesosphere. In the mesosphere, the coldest temperature is found, shown on the graph to be –90°C. In the thermosphere, the temperature continually increases and reaches very high temperatures.

Atmospheric Pressure graph – The pressure graph shows that pressure continually decreases as altitude increases. This make sense because the higher one goes, the fewer air molecules are present, producing less pressure. The *x*-axis shows that, at sea level, the air pressure is equal to one atmosphere.

Water Vapor graph – The water vapor chart shows that almost all the water vapor of the atmosphere is found in the troposphere and that the amount decreases almost to zero in the stratosphere. Because almost all the water vapor is located in the troposphere, just about all the weather is confined to this layer.

Additional Information:

- The composition of the atmosphere can be found in the Average Chemical Composition of Earth's Crust, Hydrosphere, and Troposphere section (see page 40).

- Ozone, which absorbs UV rays, is located mostly in the stratosphere. This is what causes the temperature of this layer to increase.

- The troposphere contains around 80% of the mass of the total atmosphere.

- Outer space is generally agreed to begin at 100 km or 62 miles, being in the thermosphere.

- Air pressure at sea level is equal to 14.7 pounds per square inch.

Diagram:

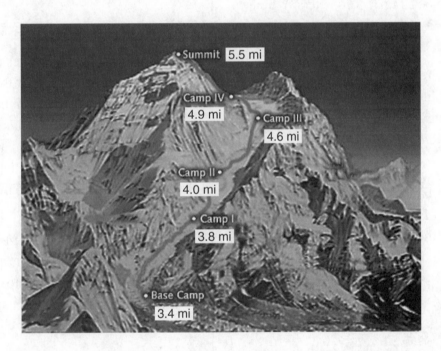

Mount Everest – The tallest mountain on our planet is Mount Everest. It is referred to as "The Top of the World" having an elevation of 29,029' or 5.5 miles. As climbers ascend, air pressure and temperature quickly deceases. At the summit, the air pressure and oxygen content is a third of that at sea level. Still, this mountain is totally within the troposphere, our first layer of the atmosphere.

1. An air temperature of 95°C most often exists in which layer of the atmosphere?

   (1) troposphere
   (2) stratosphere
   (3) mesosphere
   (4) thermosphere   1 _____

2. Most of the water vapor in the atmosphere is found in the

   (1) mesosphere
   (2) thermosphere
   (3) troposphere
   (4) stratosphere   2 _____

3. Ozone is concentrated in Earth's atmosphere at an altitude of 20 to 35 kilometers. Which atmospheric layer contains the greatest concentration of ozone?

   (1) mesosphere
   (2) thermosphere
   (3) troposphere
   (4) stratosphere   3 _____

4. As a weather balloon released from the surface of Earth rises through the troposphere, the instruments it carries will usually indicate that

   (1) temperature increases, but atmospheric pressure and concentration of water vapor decrease
   (2) temperature decreases, but atmospheric pressure and concentration of water vapor increase
   (3) temperature, atmospheric pressure, and concentration of water vapor decrease
   (4) temperature, atmospheric pressure, and concentration of water vapor increase   4 _____

5. Which graph best shows the general effect that differences in elevation above sea level have on the average annual temperature within the troposphere?

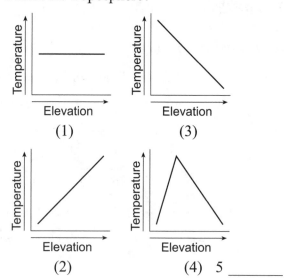

6. At an altitude of 95 miles above Earth's surface, nearly 100% of the incoming energy from the Sun can be detected. At 55 miles above Earth's surface, most incoming x-ray radiation and some incoming ultraviolet radiation can no longer be detected. This missing radiation was most likely

   (1) absorbed in the thermosphere
   (2) absorbed in the mesosphere
   (3) reflected by the stratosphere
   (4) reflected by the troposphere   6 _____

7. The movement of volcanic ash occurred at an altitude of 1.5 kilometers. State the name of the layer of Earth's atmosphere in which the ash cloud traveled.

   _____

Set 2 — Selected Properties of Earth's Atmosphere

8. In which atmospheric temperature zone does most precipitation occur?

(1) thermosphere (2) mesosphere (3) stratosphere (4) troposphere 8_______

9. Which layer of Earth's atmosphere contains very little water vapor, has an atmospheric pressure of 0.25 atmosphere, and has an air temperature that increases with altitude?

(1) troposphere (2) stratosphere (3) mesosphere (4) thermosphere 9_______

10. Which spheres are zones of Earth's atmosphere?

(1) lithosphere, hydrosphere, and troposphere
(2) stratosphere, mesosphere, and thermosphere
(3) asthenosphere, lithosphere, and hydrosphere
(4) hydrosphere, troposphere, and stratosphere 10_______

11. The graph shows the average concentration of ozone in Earth's atmosphere over Arizona during 4 months of the year. Which layer of Earth's atmosphere contains the greatest concentration of ozone?

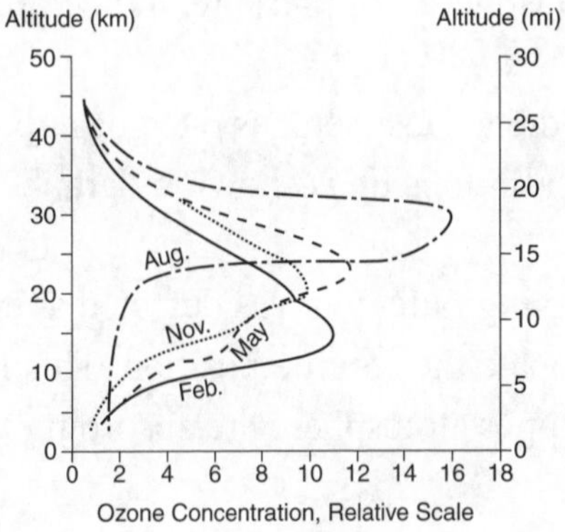

(1) troposphere
(2) stratosphere
(3) mesosphere
(4) thermosphere 11_______

12. The cross section shows the general movement of air within a portion of Earth's atmosphere located between 30° N and 30° S latitude. Numbers 1 and 2 represent different locations in the atmosphere. Which temperature zone layer of Earth's atmosphere is shown in the cross section?

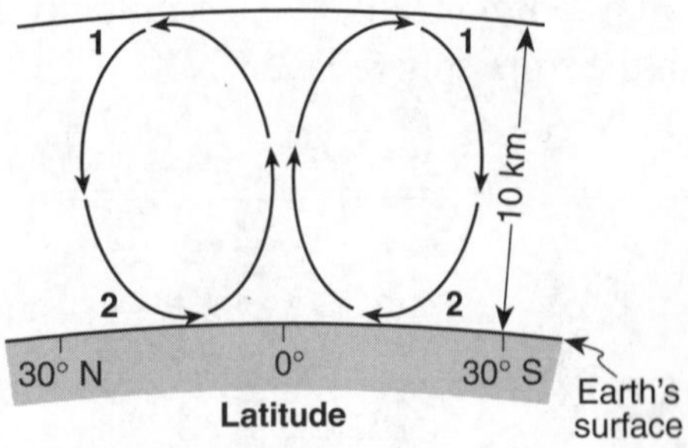

(Not drawn to scale)

(1) troposphere (3) mesosphere
(2) stratosphere (4) thermosphere 12_______

13. A mountain climber climbs to the summit of a 7 km mountain and then returns to his campsite located at the base of the mountain. Explain what happens to the temperature and pressure as he ascends to the summit and returns to his campsite.

Temperature – ____

Pressure – ____

Selected Properties of Earth's Atmosphere

Answers

Set 1

1. 4 In the Temperature Zones chart, locate 95°C on the temperature axis. From this position, move directly up to the intersection of the graph line. This position is in the thermosphere.

2. 3 In the Water Vapor chart, it shows that at the Earth's surface there is 40 g/m^3 of water vapor. At the top of the troposphere (the tropopause) there is very little water vapor. This is why almost all the weather is located in the troposphere, where the water vapor is located.

3. 4 On the altitude scale, locate the 20- to 35- km range. This range is in the stratosphere layer. Ozone absorbs much ultraviolet radiation, causing the temperature to increase within the stratosphere.

4. 3 In the troposphere, all three graph lines move toward the upper left as altitude increases. This direction is a decreasing value for all three graph variables. So, as the altitude increases within the troposphere, the pressure, temperature and water vapor decrease.

5. 3 Within the troposphere, as one goes up in elevation (altitude), the temperature decreases. This inverse relationship is shown by graph 3.

6. 1 The missing radiation was detected at an altitude of 55 miles above Earth's surface. On the altitude scale, locate the 55 mile position. This height is within the thermosphere layer. The air molecules of the upper atmosphere are absorbing this radiation.

7. Answer: Troposphere

   Explanation: 1.5 km is only about 1 mile up in the atmosphere. As shown on the altitude scale, this height is in the troposphere layer.

Planetary Wind and Moisture Belts in the Troposphere

Planetary Wind and Moisture Belts in the Troposphere

The drawing on the right shows the locations of the belts near the time of an equinox. The locations shift somewhat with the changing latitude of the Sun's vertical ray. In the Northern Hemisphere, the belts shift northward in the summer and southward in the winter.

(Not drawn to scale)

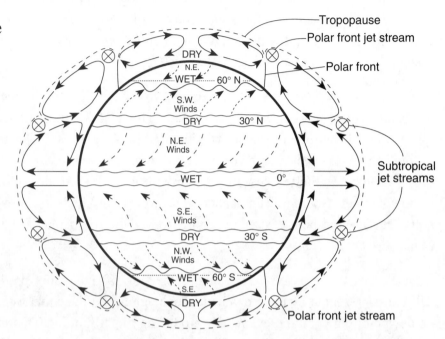

Overview:

Wind seems to blow from every direction. Yet, if one were to plot the direction of the wind over a long period of time, one direction would prevail. Winds are identified by the direction they come from. For example, we live in a wind belt in which the prevailing winds come from the west and southwest. This wind belt is labeled S.W. Winds, and known as "The Westerlies." Throughout the Earth there are six major wind belts in which the winds move in a prevailing direction. The unequal heating of our planet sets up these wind belts and are affected by the rotation of our planet. Simply, the equator area gets "overheated," causing the atmosphere to quickly heat up during the day. This heated air expands, becoming less dense, and rises into the upper levels of the troposphere where it spreads out. At the polar regions, the cold, denser air sinks and flows to low latitude regions. This process produces large convection currents within the troposphere. This, along with the Earth's rotation, sets up these large wind belts.

The moisture belts of our planet are also related to the unequal heating of our planet. At the equator, the hot rising air expands, cools, and reaches the dewpoint temperature. At this temperature, condensation of the water vapor occurs, producing massive clouds that daily release large amounts of rain, forming the major equatorial rain forests. Eventually the rising air cools and begins to slowly sink. This drier, sinking air undergoes compression that causes the air to warm up. This warm, drier air, descends on the Earth's surface around the 30° N and 30° S latitude regions, producing the many large desert and arid areas of our planet.

Wind is nothing more than the atmosphere in motion, but as you can see, the science and dynamics of wind is complicated, affecting weather and climate globally.

The Globe:

There are two groups of arrows: the ones inside the globe represent the major wind belts, and the ones outside the globe represent the major convection currents, areas of rising and sinking air cells within the troposphere, setting up the planetary wind belts. *Note:* These arrows are really one complex and dynamic system, but will be explained separately.

Arrows inside the globe – From the equator to the 30° N latitude line, the wind generally blows from the northeast. This wind-belt has been named "The Trade Winds". The next wind belt extending from the 30° N to 60° N latitude line is our wind belt, "The Westerlies". The arrows show that the prevailing winds in this belt are from the southwest. The last wind belt in the Northern Hemisphere is located between the 60° N to 90° N latitude lines. In this northern polar region, the cold, dry, prevailing winds are from the northeast. In the Southern Hemisphere three more wind belts are shown, making a total of six global wind belts. These wind belts, extending to the top of the troposphere, have a great influence on weather and climate.

Arrows outside the globe – The unequal heating of our planet produces major atmospheric convection cells, as represented by the arrows outside the globe. At the equator the hot moist rising air cools, reaching its dewpoint temperature, eventually producing much rainfall. This is noted at the equator with the word "Wet". The 60° N and 60° S latitude areas are also labeled "Wet", but overall these regions have much less rainfall when compared to the equatorial region. The descending arrows, representing sinking air, reach the Earth's surface near the 30° N and 30° S latitude lines. This drier sinking air becomes compressed and warms up, producing the major deserts of our planet in these latitude regions. These deserts and arid areas are noted on the globe by being labeled "Dry". The polar regions are also labeled "Dry". The very cold polar air holds little moisture, which makes the polar regions some of the driest areas on Earth. The globe shows two sets of jet streams, the polar front jet stream and the subtropical jet stream. These jet streams are narrow belts of high-speed wind that move weather systems.

Additional Information:

- The Coriolis effect is responsible for the curvature of wind within their respective planetary wind belt. This is caused by the rotation of our Earth.

- Planetary wind belts are responsible for moving weather systems (air masses and hurricanes). This is the reason that most of our weather comes from the west.

- Convergent wind zones are where planetary wind belts meet. See the equator zone, 0°.
  Divergent wind zones are where wind belts are moving away from each other.
  See the 30° N and 30° S area.

- Winds are caused by differences in atmospheric temperature and pressure. Winds blow from high to low-pressure areas.

- Sinking, dry air tends to produce dry, H-pressure zones – for example, 30° N and 30° S area.

- Rising, moist air tends to produce wet, L-pressure zones – for example, the equator area.

- Planetary winds produce surface ocean currents. The wind, moving in one general direction, pushes the ocean water to produce these currents (see Surface Ocean Currents map, page 60).

Planetary Wind and Moisture Belts in the Troposphere

Diagrams:

1. **Jet Streams** – The jet streams are relatively narrow bands of strong wind in the upper levels of the troposphere. In the Northern Hemisphere, the winds blow from west to east, but these belts shift southward and northward in the summer and winter respectively. These strong, prevailing westerly winds move air masses and their related fronts towards the east.

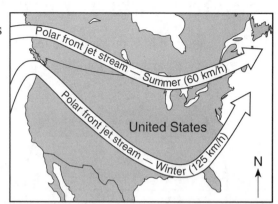

---

2. **Pressure Systems** – This L pressure system is being pushed towards the northeast by the prevailing southwest winds that are located between the 30° N and 60° N latitude lines. This is the reason why the weather to our west becomes our weather days later.

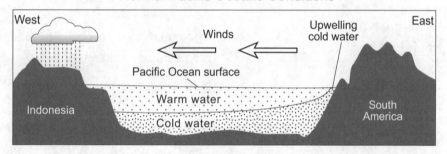

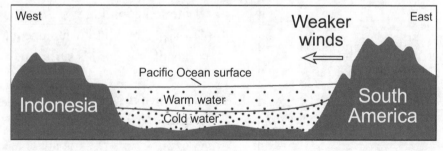

3. **El Niño** – El Niño is an abnormal weather pattern caused by the warming of the Pacific Ocean near the equator off the west coast of South America. The planetary winds in this equatorial region are the northeast and southeast wind belts that converge at the equator and blow toward the west. In an El Niño year, these planetary winds weaken, and at times have reversed, causing the warm water to accumulate along the west coast of South America. These factors produce major global weather disturbances.

---

1. The planetary winds on Earth are indicated by the curving arrows in the diagram below.

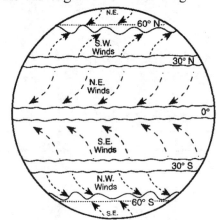

The curved paths of the planetary winds are a result of

(1) changes in humidity
(2) changes in temperature
(3) Earth's rotation on its axis
(4) Earth's gravitational force   1 _____

2. Which climate conditions are typical of regions near the North Pole and the South Pole?

(1) low temperature and low precipitation
(2) low temperature and high precipitation
(3) high temperature and low precipitation
(4) high temperature and high precipitation

2 _____

3. When the eye of a hurricane is at the 43° N latitude, it most likely is being pushed by planetary winds toward the

(1) northwest   (3) southwest
(2) northeast   (4) southeast   3 _____

4. The air over the Equator generally rises because the air is

(1) dry and cool with low density
(2) moist and hot with low density
(3) moist and cool with high density
(4) dry and hot with high density

4 _____

5. Which map best shows the surface movement of winds between 30° N and 30° S latitude?

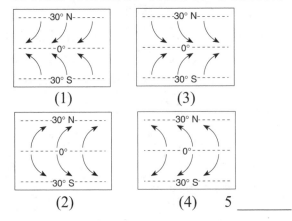

(1)   (3)

(2)   (4)   5 _____

6. The planetary wind belts in the troposphere are primarily caused by the

(1) Earth's rotation and unequal heating of Earth's surface
(2) Earth's rotation and Sun's gravitational attraction on Earth's atmosphere
(3) Earth's revolution and Sun's gravitational attraction on Earth's atmosphere
(4) Earth's revolution and unequal heating of Earth's surface   6 _____

7. A high air-pressure, dry-climate belt is located at which Earth latitude?

(1) 0°   (3) 30° N
(2) 15° N   (4) 60° N   7 _____

8. Explain why the equatorial area is the home for our planet's major rain forests?

_____

_____

9. In which map does the arrow show the general direction that most low-pressure storm systems move across New York State?

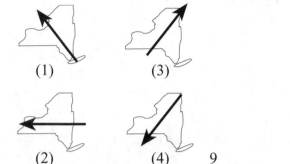

(1)   (3)

(2)   (4)   9 _____

10. The map below shows part of North America.

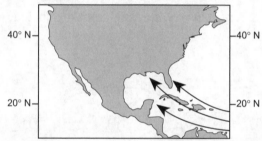

The arrows shown on the map most likely represent the direction of movement of

(1) Earth's rotation
(2) ocean conduction currents
(3) the prevailing northeast winds
(4) Atlantic Ocean hurricanes   10 _____

11. Earth's entire equatorial climate zone is generally a belt around Earth that has

(1) high air pressure and wet weather
(2) high air pressure and dry weather
(3) low air pressure and wet weather
(4) low air pressure and dry weather   11 _____

Note: Question 12 has only three choices.

12. If a parcel of air is heated, its density will

(1) decrease
(2) increase
(3) remain the same   12 _____

13. Descending air at the 30° N and S regions becomes warmer by

(1) compression of air molecules
(2) expansion of air molecules
(3) evaporation of water vapor molecules
(4) absorption of ultraviolet radiation   13 _____

14. The seasonal shifts of Earth's planetary wind and moisture belts are due to changes in the

(1) distance between Earth and the Sun
(2) amount of energy given off by the Sun
(3) latitude that receives the Sun's vertical rays
(4) rate of Earth's rotation on its axis   14 _____

15. Which graph best shows the average annual amounts of precipitation received at different latitudes on Earth?

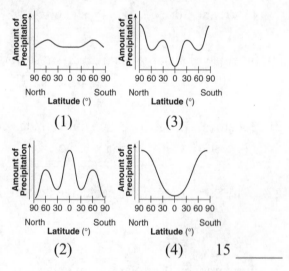

(1)   (3)

(2)   (4)   15 _____

16. Which planetary wind pattern is present in areas of great rainfall?

(1) winds diverge and air rises
(2) winds diverge and air sinks
(3) winds converge and air rises
(4) winds converge and air sinks   16 _____

Base your answers to question 17 on the diagram, which represents the planetary wind and moisture belts in Earth's Northern Hemisphere.

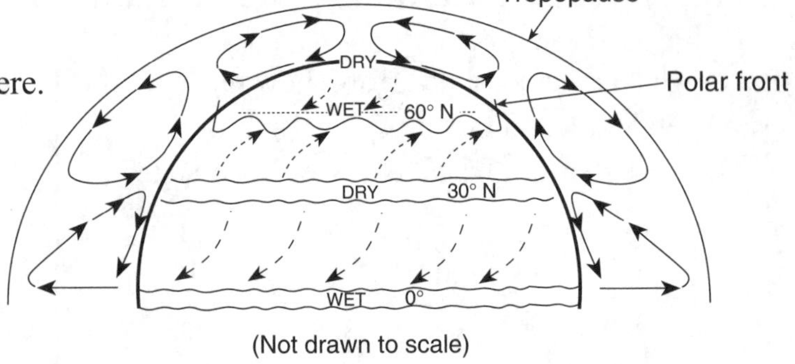

(Not drawn to scale)

17. *a)* The climate at 90° north latitude is dry because the air at that location is usually
    (1) warm and rising
    (2) warm and sinking
    (3) cool and rising
    (4) cool and sinking   a _____

    b) The paths of the surface planetary winds are curved due to Earth's
    (1) revolution   (2) rotation   (3) tilt   (4) ocean surface currents   b _____

18. The cross sections below show different patterns of air movement in Earth's atmosphere. Air temperatures at Earth's surface are indicated in each cross section. Which cross section shows the most likely pattern of air movement in Earth's atmosphere that would result from the surface air temperatures shown?

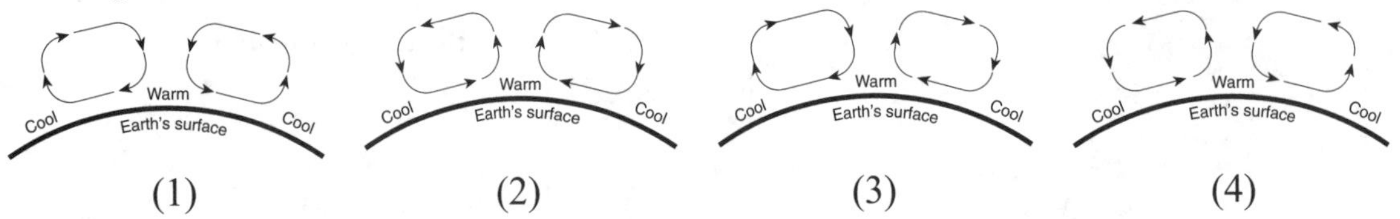

   (1)   (2)   (3)   (4)   18 _____

19. Dry areas caused by sinking air and diverging surface winds are located at which two latitudes?
    (1) 0° and 30° N   (2) 0° and 60° S   (3) 30° N and 30° S   (4) 60° N and 60° S   19 _____

20. In which planetary wind belt do most storms move toward the northeast?
    (1) 30° N to 60° N   (2) 0° to 30° N   (3) 0° to 30° S   (4) 30° S to 60° S   20 _____

21. Jet stream winds over the United States generally move from
    (1) east to west   (2) west to east   (3) north to south   (4) south to north   21 _____

22. Why do most hurricanes change direction when they pass over the 30° N latitude line?

_____

23. Explain why the polar region is labeled Dry.

_____

24. In which layer of the atmosphere would the equatorial low-pressure system of rising, warm, moist air be located in? _____

Planetary Wind and Moisture Belts in the Troposphere

1. 3 This diagram shows the major wind belts of our planet. The curving arrows represent the prevailing wind in each of the wind belts. These winds are curving (being deflected) due to the Coriolis effect that is caused by the rotation of the Earth.

2. 1 The polar regions experience the coldest temperatures of our planet, but what about the moisture? Open to the Planetary Wind and Moisture Belts chart. On this chart, the polar regions are labeled "Dry." In these polar regions, the very cold, sinking air can hold little moisture, making these areas some of the driest regions of our planet.

3. 2 Open to the Planetary Wind and Moisture Belts chart and estimate the 43° N latitude area. At this position, the prevailing wind is from the southwest. Thus, a hurricane would be pushed toward the northeast.

4. 2 In the Planetary Wind and Moisture Belts chart, locate the equator area. Shown here are ascending arrows indicating hot, rising air. Due to the Sun's strong insolation, the moist air heats up and expands, causing the density to decrease and the air to rise. This rising air cools, producing much precipitation, forming the major rain forests of our planet. This is shown on the chart with the equator area being labeled "Wet."

5. 1 The diagrams show surface winds within two wind belts near the equator. The correct direction of the wind, within each wind belt, is shown in the Planetary Wind and Moisture Belts chart. Choice 1 shows the correct directions.

6. 1 The unequal heating of our planet causes large areas of air to rise and sink. This, along with the rotation of the Earth, produces wind belts located in the troposphere.

7. 3 A high pressure zone is caused by dry, sinking air. The Planetary Wind chart shows air sinking at the 30° N and S regions. This dry, descending air undergoes compression causing its temperature to increase. In these latitude belts, the Earth's major deserts are located.

8. The direct sunlight on the equator area causes the air to quickly heat up during the day. The hot air expands and rises. The rising air cools and eventually reaches the dewpoint temperature, forming large rain clouds. This process repeats itself almost daily, producing precipitation in the major rain forests of our planet.

Electromagnetic Spectrum

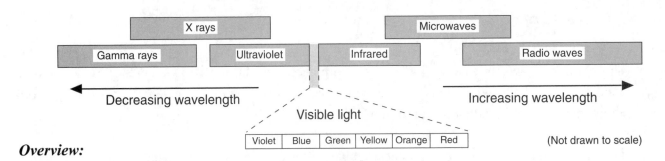

Violet | Blue | Green | Yellow | Orange | Red

(Not drawn to scale)

Overview:

The electromagnetic (EM) spectrum consists of waves with both electric and magnetic properties. All electromagnetic waves travel at the speed of light. The difference between them is their wavelengths. Gamma rays have the shortest wavelengths, and radio waves have the longest wavelengths. The shortest waves, gamma rays and x-rays, are dangerous due to their penetrating power. The ultraviolet (UV) rays are those harmful sun-tanning waves. The visible light that is being reflected off this paper is the only part of the spectrum that we can see. The infrared waves can not be seen, but are felt as heat. The Sun radiates the total electromagnetic spectrum continuously, but radiates more visible waves than any other part of the spectrum. The Earth absorbs these waves during the day and heats up. At night, the Earth reradiates much of this energy back to space in the form of infrared waves – heat waves. Some of the infrared waves are absorbed by greenhouse gases, maintaining the Earth's atmospheric temperature. This process is needed, giving our planet a temperature that is favorable for life. But, with the increase of greenhouse gases, methane, water vapor, and especially CO_2 that is released by burning fossil fuel, more infrared waves are being absorbed. This is causing an increase in global temperature. The consequences of this situation are being documented globally, causing more severe weather, gradual climate changes and increase of extinction risk for certain species.

The Chart:

Visible light, the only part of the electromagnetic spectrum that we can see, has been expanded to show the relative wavelengths of the six colors that produce white light. Violet, being on the left side, has the shortest wavelength and red, being on the right side, has the longest wavelength. Moving left from visible light, as the arrow indicates, are electromagnetic waves that are decreasing in wavelength. Moving to the right of visible light are the longer wavelengths with radio waves being the longest of all. Some of these wavelengths, especially microwaves and x-rays, overlap others, as shown by their gray bars.

Additional Information:

- When an object absorbs electromagnetic energy, it will later reradiate energy in longer wavelengths.

- Smooth, light/shiny surfaces reflect much visible radiation, while dark, rough surfaces absorb much radiation.

- Ozone, located mostly in the stratosphere, absorbs much harmful ultraviolet (UV) rays.

- The word insolation is a combination of 3 words: incoming, solar, and radiation.

1. Which process transfers energy primarily by electromagnetic waves?

   (1) radiation   (3) conduction
   (2) evaporation   (4) convection   1 _____

2. Which type of radiation has the shortest wavelength?

   (1) microwaves   (3) ultraviolet
   (2) visible light   (4) infrared   2 _____

3. Of the following choices, which visible radiation has the shortest wavelength?

   (1) red   (3) green
   (2) blue   (4) yellow   3 _____

4. Which list contains three major greenhouse gases found in Earth's atmosphere?

   (1) carbon dioxide, methane, and water vapor
   (2) carbon dioxide, oxygen, and nitrogen
   (3) hydrogen, oxygen, and methane
   (4) hydrogen, water vapor, and nitrogen

                             4 _____

5. In which list are the forms of electromagnetic energy arranged in order from longest to shortest wavelengths?

   (1) gamma rays, x-rays, ultraviolet rays, visible light
   (2) radio waves, infrared rays, visible light, ultraviolet rays
   (3) x-rays, infrared rays, blue light, gamma rays
   (4) infrared rays, radio waves, blue light, red light   5 _____

6. Most of the energy radiated by Earth's surface at night is in the form of

   (1) infrared rays   (3) visible light rays
   (2) ultraviolet rays   (4) x-rays   6 _____

7. Compared to dull and rough rock surfaces, shiny and smooth rock surfaces are most likely to cause sunlight to be

   (1) reflected   (3) scattered
   (2) refracted   (4) absorbed   7 _____

8. Which diagram best represents the wavelength of most of the sunlight energy absorbed and the wavelength of infrared energy reradiated by the roof of a building at 2 p.m. on a clear summer day?

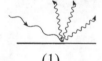   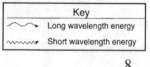

    (1)   (2)   (3)   (4)   8 _____

9. Which graph best represents the relative wavelengths of the different forms of electromagnetic energy?

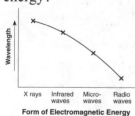

    (1)   (2)   (3)   (4)   9 _____

10. Which form of electromagnetic radiation can be shorter than ultraviolet, but longer than gamma rays?

    (1) microwaves   (3) infrared
    (2) ultraviolet   (4) x-rays   10_____

11. In the visible spectrum, which color has the longest wavelength?

    (1) red   (3) orange
    (2) green   (4) violet   11_____

12. What part of the Sun's electromagnetic spectrum has wavelengths between ultraviolet and infrared radiation?

    (1) radio waves radiation
    (2) microwaves radiation
    (3) visible radiation
    (4) x-ray radiation   12_____

13. What is the basic difference between ultraviolet, visible, and infrared radiation?

    (1) half-life   (3) wavelength
    (2) temperature   (4) wave velocity 13_____

14. Which process is responsible for the greatest loss of energy from Earth's surface into space on a clear night?

    (1) condensation   (3) radiation
    (2) conduction   (4) convection 14_____

15. Equal areas of which type of surface will reflect the most insolation?

    (1) light gray rooftop
    (2) dark tropical forest
    (3) snow-covered field
    (4) black paved road   15_____

16. Which diagram best represents visible light rays after striking a dark, rough surface?

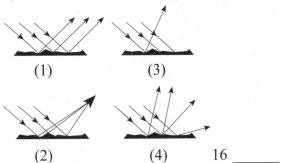

    (1)   (3)

    (2)   (4)   16_____

17. Scientists have theorized that an increased concentration of carbon dioxide will cause an increase in worldwide atmospheric temperature. This theory is based on the fact that carbon dioxide is a

    (1) good absorber of infrared radiation
    (2) poor absorber of infrared radiation
    (3) good reflector of ultraviolet radiation
    (4) poor reflector of ultraviolet radiation

                                17_____

18. Scientists are concerned about the decrease in ozone in the upper atmosphere primarily because ozone protects life on Earth by absorbing certain wavelengths of

    (1) x-ray radiation
    (2) ultraviolet radiation
    (3) infrared radiation
    (4) microwave radiation   18_____

19. If average wavelength of infrared radiation is 10^{-3} cm. Which of the wavelengths below could represent the average wavelength of microwaves?

    (1) 10^{-1} cm   (3) 10^{-7} cm
    (2) 10^{-4} cm   (4) 10^{-10} cm   19_____

20. Energy is transferred from the Sun to Earth mainly by

(1) molecular collisions
(2) density currents
(3) electromagnetic waves
(4) red shifts

20 _____

21. Cosmic microwave background radiation is classified as a form of electromagnetic energy because it

(1) travels in waves through space
(2) is visible to humans
(3) moves faster than the speed of light
(4) moves due to particle collisions

21 _____

22. The diagram below shows the types of electromagnetic energy given off by the Sun. The shaded part of the diagram shows the approximate amount of each type actually reaching Earth's surface.

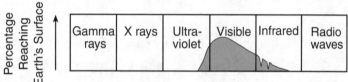

Which conclusion is best supported by the diagram?

(1) All types of electromagnetic energy reach Earth's surface.
(2) Gamma rays and *x*-rays make up the greatest amount of electromagnetic energy reaching Earth's surface.
(3) Visible light makes up the greatest amount of electromagnetic energy reaching Earth's surface.
(4) Ultraviolet and infrared radiation make up the greatest amount of electromagnetic energy reaching Earth's surface.

22_____

23. Which diagram best represents how greenhouse gases in our atmosphere trap heat energy?

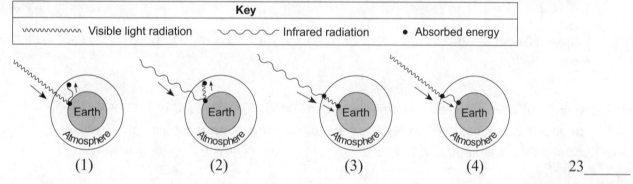

23_____

24. What physical property is different between visible waves and radio waves?

_____

25. Describe two human activities that would decrease the amount of carbon dioxide that humans add to the Earth's atmosphere. 1_____ 2_____

1. 1 Electromagnetic waves are transferred by the process of radiation. The Sun radiates all of the different types of electromagnetic waves. Conduction is the transfer of heat in solids. Convection is the transfer of heat in liquids and gases.

2. 3 Open to the Electromagnetic Spectrum (EM) chart. Of the given choices, ultraviolet has the shortest wavelength. All of the other answers are to the right of the ultraviolet wavelength section, making them longer in wavelengths.

3. 2 In the EM Spectrum chart, the visible light is expanded to show the wavelengths of the different colors. Using this expanded chart, blue wavelengths are the shortest of the given choices.

4. 1 CO_2, methane, and water vapor are considered greenhouse gases. These gases are able to absorb infrared rays radiated by the Earth. As more greenhouse gases are released into the atmosphere, more infrared radiation is absorbed resulting in higher atmospheric temperatures.

5. 2 Radio waves are the longest waves. From this position, on the chart, as you move to the left, the waves decrease in length.

6. 1 During the day the Earth absorbs mostly visible and ultraviolet radiation. At night, the Earth reradiates longer wavelengths in the infrared part of the spectrum.

7. 1 Shiny and smooth surfaces reflect much radiation, while darker, rougher surfaces tend to absorb more radiation.

8. 4 A roof will absorb the Sun's short wavelengths (mostly visible) and warm up. At a certain temperature, the roof gets too hot and starts reradiating longer wavelengths. This usually occurs in the afternoon on a hot, sunny day. Diagram 4 correctly shows this situation.

9. 2 Open to the EM Spectrum chart. The shortest wavelengths are located on the far left side; they are gamma rays and X-rays. As you move to the right, the waves become increasingly longer, with radio waves being the longest.

Electromagnetic Spectrum

Characteristics of Stars

(Name in italics refers to star represented by a ⊕.)
(Stages indicate the general sequence of star development.)

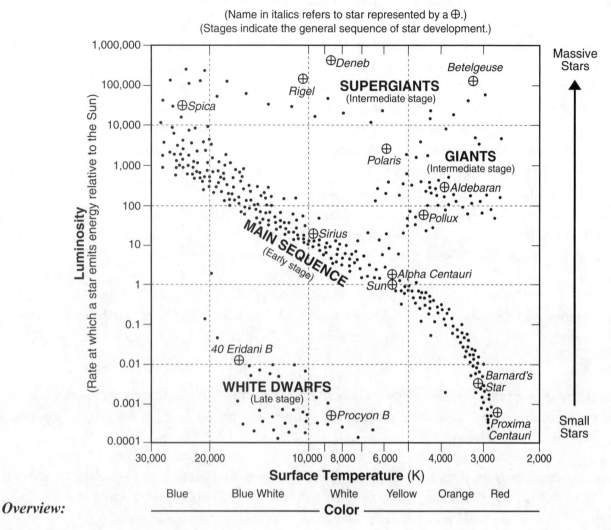

Overview:

When the temperature and luminosity (brightness) of stars were first plotted on a graph, a pattern was observed. Most of the stars fell within a specific region that ran diagonally across the graph, later to be named the main sequence. Stars off the main sequence fell mainly into groups of giants, supergiants, and white dwarfs. Years later, it was discovered that this pattern revealed stellar evolution according to the position of a star on this graph. The two scientists that independently produced this chart, Hertzsprung and Russell, are credited for making a tremendous contribution to astronomy. Thus, this graph is referred to as the H-R diagram.

Stars, through nuclear fusion, convert mostly hydrogen into the heavier element, helium. This nuclear reaction continues for millions or even billions of years, generating the star's energy. Eventually, when the hydrogen fuel starts to become exhausted, a star, in reaction to the force of gravity within its core, expands greatly, becoming a cooler giant or supergiant star. In time, these dying giant stars either collapse, becoming a white dwarf star, or will undergo a supernova explosion – a violent stellar explosion.

Our Sun is an average yellow star and is presently positioned on the main sequence. When it enters its intermediate stage, the Sun will greatly expand becoming a red giant star. This will vaporize some of the inner planets, while toasting the Earth. The Sun will eventually collapse becoming a hot white dwarf star. But not to worry, this will not happen this year, so concentrate on passing the regents.

The Graph:

The *x*-axis is the Surface Temperature (K) of stars. The surface temperature of a star produces its color. Red stars are the coolest, having a temperature around 2,600 K, and blue stars are the hottest having temperatures over 20,000 K. Our Sun has a surface temperature of just under 6,000 K, making it a yellow star. The *y*-axis is the Luminosity scale. This scale represents the relative brightness of a star compared to our Sun, if that star and our Sun were placed side by side at a given distance from the Earth. Our Sun is assigned the luminosity value of 1. Brighter stars have a luminosity value higher than 1, and duller stars have a value less than 1. On the far right side is a descriptive scale showing the size of stars from "Small Stars" to "Massive Stars." As shown on the diagram, small stars have low luminosity values and would be dull stars in the night sky. Massive stars, having very high luminosity values, are bright stars in the night sky. For example, 40 Eridani B is a white dwarf star, shown to be quite hot, but being small, it is relatively dull, as shown by its low luminosity value. Betelgeuse is a red supergiant that has a relatively cool temperature of around 3,500 K, but due to its large size (a true monster of a star), Betelgeuse has a luminosity 100,000 times that of our Sun. This red star is easily seen in the winter constellation Orion.

Early Stage – As mentioned in the introduction, stars go through different stages. For most of their lives, they are positioned on the main sequence. Their position on the main sequence is related to the mass of the star. Throughout this extended early stage, nuclear fusion converts hydrogen into the heavier element of helium.

Intermediate Stage – Eventually, as the star's hydrogen fuel becomes limited, it greatly expands, causing its surface temperature to cool. This will reposition the star off the main sequence to the giant or supergiant position. It has now entered into its intermediate stage – its dying stage.

Late Stage – These now dying, giant stars, depending on their mass, will either under a supernova explosion, or contract becoming a very dense, hot, white dwarf stars. These white dwarf stars, even with their relatively high temperatures, are so small they all have low luminosity values.

Additional Information:

- The Sun's energy is the result of nuclear fusion – an energy producing process. The equation for the fusion reaction of hydrogen is:

$$\text{Hydrogen} + \text{Hydrogen} \rightarrow \text{Helium} + \text{Energy}$$
    (lighter element)   (lighter element)   (heavier element)

- Our Sun is estimated to be 4.6 billion years old. Our universe is estimated to be 13 to 14 billion years old.

- The Big Bang Theory states that extremely hot cosmic radiation was released with the "big bang explosion" that created our universe. With the expansion of the universe, the temperature of the cosmic radiation cooled significantly. Astronomers have detected and measured this remaining cosmic radiation providing strong supporting evidence for the origin and age of our universe.

- Astronomers believe that the collapse of the core of a massive star forms a black hole. This black hole creates so much gravity that light can't escape. It is believed that most, if not all galaxies, rotate around a centrally located supermassive black hole.

Characteristics of Stars

Diagrams:

1. **The Big Bang Explosion** – According to the Big Bang theory, astronomers estimate that our universe began 13.6 billion years ago. Immediately following this massive explosion, energy and elementary matter expanded outward producing our expanding universe. In time, due to the force of gravity, interstellar clouds of gases and dust contracted producing stars within galaxies as shown by letter *A*. Proofs of the Big Bang theory are the leftover measurable cosmic radiation from the initial explosion and the redshift of spectral lines of distant celestial objects that show these objects are still moving away from us at tremendous speeds.

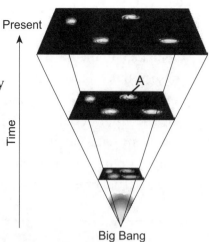

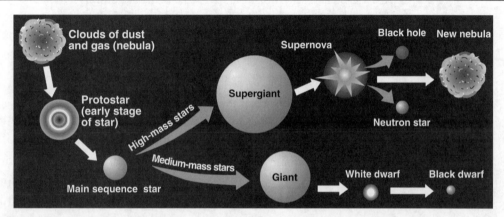

2. **Stages of Star Evolution** – The diagram shows the possible sequences in the life cycle of stars, beginning with their formation from nebula gas clouds in space to their dying stage. When nuclear fusion occurs, where lighter nuclei join to produce a heavier nucleus, a star is born. It radiates energy and takes its place on the main sequence. The life cycle followed by a star is determined by the star's mass. As the star's fuel becomes limited, a star like our Sun will eventually expand, move off the main sequence, collapse, and become a white dwarf. More massive stars will take other paths. Throughout this complete cycle, the force of gravity is involved in all events.

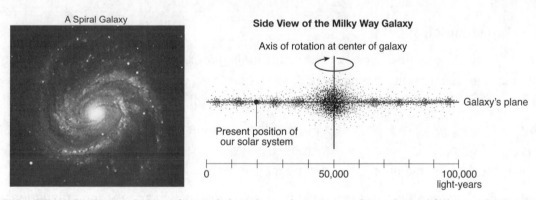

3. **The Milky Way Galaxy** – It is estimated that the universe consists of 100 billion galaxies. Our Milky Way galaxy may contain close to 200 billion stars, our Sun being one of them. The shape of our galaxy is spiral, with our solar system located on an outer arm of our rotating galaxy. It is theorized that galaxies rotate due to a supermassive black hole located at the center of all galaxies.

4. **Redshift of Spectral Lines** – The spectral lines of hydrogen gas from two galaxies, *A* and *B*, are compared to the spectral lines of hydrogen gas observed in a laboratory. Notice that both galaxies' spectral lines have shifted toward the red end of the spectrum. A redshift of spectral lines occurs when any celestial object is moving away from Earth. A blueshift of spectral lines occurs when any celestial object is moving towards Earth.

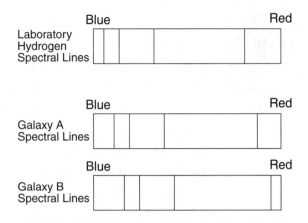

Stars	Temperature		Luminosity	
	Hotter	Cooler	Brighter	Dimmer
Procyon B	X			X
Barnard's Star		X		X
Rigel	X		X	

5. **Characteristics of Stars** – This chart shows the comparison of temperature and luminosity of three stars to our Sun. The information needed for this table was found in the Characteristic of Stars chart. Massive stars located on the main sequence have greater luminosity and hotter temperatures than our Sun. Small stars on the main sequence tend to be cooler with lower luminosity values. Supergiant and giant stars, off the main sequence, have greater luminosity but most are cooler than our Sun.

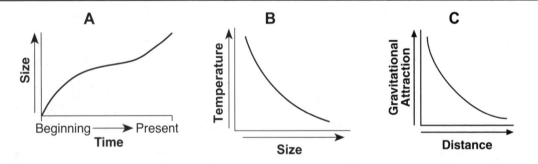

6. **Graphical Relationships:**

Graph A – According to the Big Bang theory, this graph represents the relationship between the size of the universe and time from the beginning of the universe to the present. Immediately after the Big Bang, energy expanded outward forming matter. Our universe is still expanding.

Graph B – This graph shows the relationship of the size of the universe to its temperature. Simply, as the universe expands, the temperature drops. This was proven by measuring the leftover cosmic microwave background radiation generated by the Big Bang.

Graph C – The relationship shown on this graph is: As distance increases, the gravitational attraction between two objects will decrease. It can also be stated as: As distance decreases, the gravitational attraction between two objects will increase. Gravitational attraction of cosmic dust and gases formed the stars of our galaxies.

1. Which star color indicates the hottest star surface temperature?

   (1) blue   (3) yellow
   (2) white   (4) red   1 _____

2. The graph represents the brightness and temperature of stars visible from Earth.

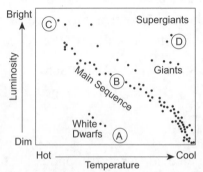

   Which location on the graph best represents a star with average brightness and temperature?

   (1) A   (2) B   (3) C   (4) D   2 _____

3. Which star is more massive than our Sun, but has a lower surface temperature?

   (1) *40 Eridani B*   (3) *Aldebaran*
   (2) *Barnard's Star*   (4) *Sirius*   3 _____

4. Which star is hotter and many times brighter than Earth's Sun?

   (1) Barnard's Star   (3) Rigel
   (2) Betelgeuse   (4) Pollux   4 _____

5. Compared with our Sun, the star Betelgeuse is

   (1) smaller, hotter, and less luminous
   (2) smaller, cooler, and more luminous
   (3) larger, hotter, and less luminous
   (4) larger, cooler, and more luminous

                                          5 _____

6. Which two stars have the most similar luminosity and temperature?

   (1) Betelgeuse and Barnard's Star
   (2) Rigel and Betelgeuse
   (3) Alpha Centauri and the Sun
   (4) Sirius and Procyon B   6 _____

7. Compared to the temperature and luminosity of the star Polaris, the star Sirius is

   (1) hotter and more luminous
   (2) hotter and less luminous
   (3) cooler and more luminous
   (4) cooler and less luminous   7 _____

8. In nuclear fusion what occurs?

   (1) Lighter elements are converted to heavier elements.
   (2) Lighter elements are converted to even lighter elements.
   (3) Heavier elements are converted to lighter elements.
   (4) Heavier elements chemically combine with lighter elements.   8 _____

9. Betelgeuse and Aldebaran are both red-giant stars. Give a statement comparing their luminosity and temperature values.

_____

10. A star located off the main sequence indicates what?

_____

11. Compared to the surface temperature and luminosity of massive stars in the Main Sequence, the smaller stars in the Main Sequence are

    (1) hotter and less luminous
    (2) hotter and more luminous
    (3) cooler and less luminous
    (4) cooler and more luminous   11 _____

12. The Sun is inferred to be the most luminous when it is classified as a

    (1) white dwarf star
    (2) gas cloud (nebula)
    (3) main sequence star
    (4) giant star   12 _____

13. For other stars in our galaxy that go through a similar life cycle to our Sun, which star is currently in the late stage of its life cycle?

    (1) Alpha Centauri   (3) Barnard's Star
    (2) Procyon B   (4) Polaris   13 _____

14. Which star is hotter, but less luminous, than Polaris?

    (1) Deneb   (3) Procyon B
    (2) Aldebaran   (4) Pollux   14 _____

15. The Sun revolves around the center of

    (1) *Polaris*
    (2) *Aldebaran*
    (3) Earth
    (4) the Milky Way Galaxy   15 _____

16. Which force is mostly responsible for the contraction of a nebula gas cloud in the formation of a new star?

    (1) friction   (3) magnetism
    (2) gravity   (4) inertia   16 _____

17. Which star's surface temperature is closest to the temperature at the boundary between Earth's mantle and core? (See page 120.)

    (1) Sirius   (3) the Sun
    (2) Rigel   (4) Betelgeuse   17 _____

18. Which evidence best supports the theory that the universe began with a massive explosion?

    (1) cosmic background radiation in space
    (2) parallelism of planetary axes
    (3) radioactive dating of Earth's bedrock
    (4) life cycle of stars   18 _____

19. Which object in space emits light because it releases energy produced by nuclear fusion?

    (1) Earth's Moon   (3) Venus
    (2) Halley's comet   (4) Polaris   19 _____

20. Great amounts of energy are released in the core of a star as lighter elements combine and form heavier elements during the process of

    (1) compaction
    (2) condensation
    (3) radioactive decay
    (4) nuclear fusion   20 _____

21. Compared to the luminosity and surface temperature of red main sequence stars, blue supergiants are

(1) less luminous and have a lower surface temperature
(2) less luminous and have a higher surface temperature
(3) more luminous and have a lower surface temperature
(4) more luminous and have a higher surface temperature   21_____

22. Light from distant galaxies most likely shows a

(1) redshift, indicating that the universe is expanding
(2) redshift, indicating that the universe is contracting
(3) blueshift, indicating that the universe is expanding
(4) blueshift, indicating that the universe is contracting   22_____

23. Explain why a giant star that is cooler than our Sun, similar to Aldebaran, has a greater luminosity

than the Sun. _____

The star chart shows part of the winter sky visible from New York State. Some of the brighter stars are labeled and the constellation Orion is outlined.

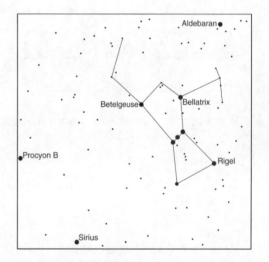

24. *a)* Identify the color of the star Bellatrix, which has a surface temperature of approximately 21,000 K.

_____

b) Excluding the star Bellatrix, which star, given on the chart, would be considered the youngest?

_____

c) In the accompanying chart, list the stars, other than Bellatrix, found on the chart in order of decreasing luminosity. Rigel, the most luminous star, has been listed.

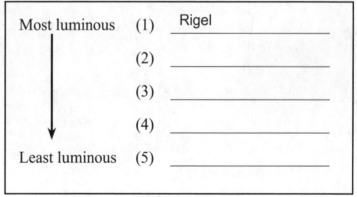

Most luminous   (1) ___Rigel_____

                  (2) _____

                  (3) _____

                  (4) _____

Least luminous   (5) _____

25. Give a statement on the relationship between temperature and luminosity of the main sequence stars.

_____

Base your answers to question 26 on the graph below and on the "Characteristics of Stars" graph. The graph below shows the inferred stages of development of the Sun, showing luminosity and surface temperature at various stages.

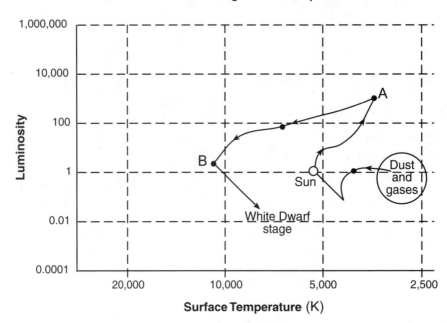

26. *a*) Describe the changes in luminosity of the Sun that will occur from its current Main Sequence stage to its final White Dwarf stage.

_____

b) What are the initial substances that were needed to form the Sun?_____

c) Identify the process that produces the Sun's energy. _____

d) Why does the Sun's luminosity increases and its temperature decreases at position *A*?

_____

e) At position *A*, what color star would be observed? _____

f) Which star shown on the "Characteristics of Stars" graph (page 189) is currently at the Sun's final predicted stage of development?_____

27. List the following astronomical features, in order of relative size, from smallest to largest.

Sun	Smallest _____
Deneb	_____
Milky Way Galaxy	_____
Universe	_____
Proxima Centauri	_____
Pollux	Largest _____

1. 1   In the Characteristics of Stars chart, go to the Temperature axis. Blue stars have the hottest surface temperatures, over 20,000 K.

2. 2   An average star would be located on the main sequence and have the same luminosity and temperature as our Sun. Locate our Sun on the graph. Location *B* is close to where our Sun is positioned.

3. 3   On the Characteristics of Stars chart, Aldebaran is classified as a giant star, which is more massive than our Sun. Aldebaran surface temperature is close to 4,000 K, which is cooler than our Sun.

4. 3   Locate Rigel on the Characteristics of Stars chart. As shown, the luminosity of this star is over 100,000 times that of our Sun. Its temperature is around 10,000 K, while the Sun's temperature is just under 6,000 K.

5. 4   Betelgeuse has a temperature around 3,200 K, which is cooler than our Sun. It has a greater luminosity than our Sun because it is a supergiant star.

6. 3   Alpha Centauri and the Sun are next to each other on the Characteristics of Stars chart. Thus, they have almost the same temperature and luminosity values.

7. 2   Sirius is positioned lower on the Luminosity scale than Polaris. This makes Sirius less luminous than Polaris. The temperature of Sirius is greater than that of Polaris, as shown by the temperature scale.

8. 1   By the process of nuclear fusion, lighter elements are joined to produce heavier element(s), producing much energy.

9. Betelgeuse has a higher luminosity and slightly lower temperature than Aldebaran.
   or
   Aldebaran has a lower luminosity and slightly higher temperature than Betelgeuse.

10. In stellar evolution when a star is off the main sequence it may indicate that:
    a) it is in its dying stage.
    b) its fuel is running out.
    c) it is expanding causing its temperature to cool.
    d) it has collapsed and is a white dwarf.
    e) it has entered the intermediate or late stage of its existence.

Solar System Data

Celestial Object	Mean Distance from Sun (million km)	Period of Revolution (d=days) (y=years)	Period of Rotation at Equator	Eccentricity of Orbit	Equatorial Diameter (km)	Mass (Earth = 1)	Density (g/cm³)
SUN	—	—	27 d	—	1,392,000	333,000.00	1.4
MERCURY	57.9	88 d	59 d	0.206	4,879	0.06	5.4
VENUS	108.2	224.7 d	243 d	0.007	12,104	0.82	5.2
EARTH	149.6	365.26 d	23 h 56 min 4 s	0.017	12,756	1.00	5.5
MARS	227.9	687 d	24 h 37 min 23 s	0.093	6,794	0.11	3.9
JUPITER	778.4	11.9 y	9 h 50 min 30 s	0.048	142,984	317.83	1.3
SATURN	1,426.7	29.5 y	10 h 14 min	0.054	120,536	95.16	0.7
URANUS	2,871.0	84.0 y	17 h 14 min	0.047	51,118	14.54	1.3
NEPTUNE	4,498.3	164.8 y	16 h	0.009	49,528	17.15	1.8
EARTH'S MOON	149.6 (0.386 from Earth)	27.3 d	27.3 d	0.055	3,476	0.01	3.3

Overview:

The Milky Way Galaxy is the home of hundreds of billions of stars, of which our solar system constitutes a very, very small part. Our solar system is mostly empty space. In the center of this "seemingly empty space" is the most massive body in our solar system – our Sun, containing more than 99% of the total mass of our solar system. Due to the Sun's large mass that creates so much gravity, all objects in our solar system revolve around the Sun. These objects, having eccentric orbits, consist of planets, minor planets, asteroids, comets, dust, gases, and other numerous objects. Our Moon is our only natural satellite. The Apollo moon program was a USA space-flight mission to successfully land a man on the Moon. This was accomplished 6 times. No other nation has landed a person on the Moon.

The Chart:

This chart gives specific data dealing with measurements and motions of our Sun, the eight planets and our Moon. Many answers can be obtained by finding the correct column and the corresponding object. Let's review some of the information and relationships that are given in this chart. The Period of Revolution column shows a relationship that the farther a planet is from the Sun, the longer the period of revolution is for that planet. This relationship exists because, as the distance from the Sun increases, the Sun's gravitational attraction decreases. The Period of Rotation column shows the length of the planet's day. Notice that Jupiter has the shortest period of rotation, making its day just under 10 hours, while Venus has the longest day of any planet. In fact, Venus' period of rotation is longer than its period of revolution, which makes Venus' day longer than its year.

The Eccentricity column is extensively covered by the Eccentricity Equation section (see page 18). Since all planets have an eccentricity value greater than zero, they all orbit the Sun in elliptical orbits. Remember, the lower the eccentricity value, the more circular the orbit will be. For the Equatorial Diameter

Solar System Data

column, you might have to use the ratio of these diameter measurements to compare two or more diagrammed circles representing the planets' sizes. For example, circles representing Venus and Earth will be almost the same size, while a diagram of Mars will be half the size of one representing the Earth. The Mass column gives the Earth's mass as 1. All other masses are compared to this standard. In the Density column, an interesting fact is revealed. Saturn, one of the giant gaseous planets, has a density less than 1. This planet would float in water, assuming one could find a body of water large enough to place it in. Notice that the terrestrial planets (also known as the rocky planets – Mercury, Venus, Earth, and Mars), as expected, have a higher density than the giant gaseous *Jovian* planets (Jupiter, Saturn, Uranus, and Neptune). The bottom row, Earth's Moon, gives information about our only natural satellite.

Earth's Moon – The Moon revolves around the Earth in an elliptical geocentric orbit (Earth-centered) and travels with the Earth around the Sun in a heliocentric orbit (Sun-centered). The Moon's period of revolution and period of rotation are the same, 27.3 d. That is why the same side of the Moon is always seen from the Earth. But, because the Earth is traveling in its orbit, it takes the Moon close to 29.5 d to complete one full cycle of Moon phases (e.g. New Moon to New Moon – a lunar month). The moon phases are caused by the revolution of the Moon around the Earth. It is the changing angle of the Moon, Earth, and Sun which produces the different amounts of reflected light off the Moon that we see. During one month, an observer on Earth would see eight different phases of the Moon: New Moon, Waxing Crescent, First Quarter, Waxing Gibbous, Full Moon, Waning Gibbous, Third Quarter, Waning Crescent. There are two different eclipses. In a solar eclipse, the New Moon moves in front of the Sun; this occurs during the day. In a lunar eclipse, the Full Moon enters the Earth's shadow; this occurs during the night. The Moon has a much stronger influence on tides than the Sun. This is because the Moon is much closer to the Earth than the Sun.

Additional Information:

- The Earth revolves 1° per day in its orbit, while rotating 15° per hour.

- In the waxing phase of the Moon, the illuminated area is increasing. This occurs after the New Moon and before the Full Moon phase. In the waning phase of the Moon, the illuminated area is decreasing. This occurs after the Full Moon and before the New Moon phase.

- The asteroid belt, consisting of thousands of large rocks, is located between Mars and Jupiter.

- Venus is the hottest planet due to its dense carbon dioxide atmosphere, producing a runaway greenhouse effect.

- Perigee and Apogee – Perigee is a point at which an object in orbit has its closest approach to the body it is orbiting. Apogee is a point at which an object in orbit is farthest from the the body it is orbiting. At perigee, the object will have its greatest orbital speed. At apogee, the orbiting object will have its slowest orbital speed.

- The apparent diameter is the size of the celestial object as seen from the Earth. Our Sun, planets, and our Moon appear to change in size as its orbital distance from the Earth increases or decreases. This is very evident by an apparently large Full Moon or "supermoon" when it is in its perigee position.

Diagrams:

1. **Our Solar System** – Although these orbits look round, they are not. All planets orbit the Sun in elliptical orbits that are slightly flattened, giving them an oval shape. Due to this orbital shape, during its period of revolution, each planet at one time will be closer to the Sun producing its greatest orbital velocity and at another time farther from the Sun, producing its slowest orbital velocity. These constant changes in orbital velocity of the planets are controlled by the gravitational force of the Sun.

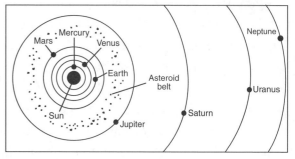

2. **Our Planets** – This diagram shows the relative sizes of the planets in our solar system. The first 4 planets (Mercury, Venus, Earth, and Mars) are the terrestrial, inner rocky planets. The outer 4 planets (Jupiter, Saturn, Uranus, and Neptune) are the giant gaseous Jovian planets. The inner planets, being rocky in composition, are denser than the gaseous planets.

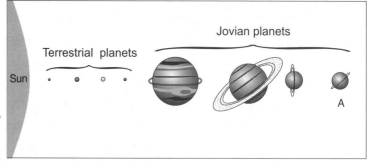

3. **Star Trails** – This is a time-exposure photograph of star trails. The trails show the apparent movement of the stars, but this perceived motion is really caused by the rotating Earth – rotating 15°/hour. These trails show an arc of 120°, which represents 8 hours of rotation. The celestial object in the center is Polaris – the North Star. Star trails are an accepted proof of Earth's rotation.

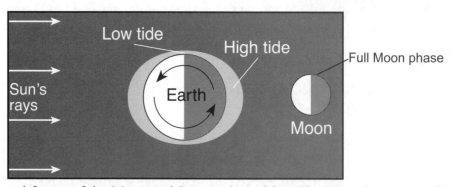

4. **Tides** – The gravitational forces of the Moon and Sun produce tides. The Moon has more of an effect on tides than the Sun because the Moon is closer to the Earth. Because of Earth's rotation, coastal locations would see two high tides and two low tides in just over 24 hours. The greatest of these high tides occurs on the side of the Earth closest to the Moon. The highest recorded tides occur when there is a New Moon. This high tide is referred to as the spring tide, but has nothing to do with seasons. Much coastal flooding can occur when a spring tide coincides with a coastal storm.

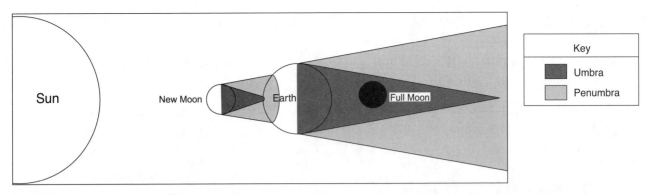

5. **Eclipses** – A solar eclipse occurs when the New Moon – the Moon that we cannot see since it is up in the sky during the day – passes in front of the Sun. The New Moon produces two shadows, with the darker shadow (umbra) producing a total solar eclipse. A lunar eclipse occurs when the Full Moon enters the Earth's shadow. The viewing of a lunar eclipse is common, unlike a total solar eclipse which is rarer. For an eclipse to occur, the Earth, Sun, and Moon must be in the same orbital plane, and this does not occur each month.

6. **Solar Eclipse** – The middle picture of this multiple-exposure photograph shows a total solar eclipse. This eclipse occurs only when the New Moon's orbit is in the proper plane with the Sun and Earth. When the New Moon moves in front of the Sun, blocking it, a solar eclipse occurs, and the Sun's corona becomes visible to Earth observers who are in the solar eclipse totality path (the umbra).

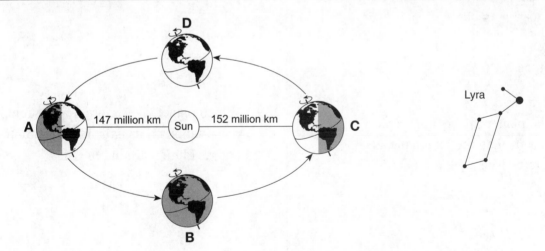

7. **View from Space** – This diagram shows the Earth's axis extending out from the poles, which is always parallel to other axis' positions in the Earth's orbit. This causes our northern axis to always point to the North Star, Polaris. One can see that during our summer (*C*), the North Pole experiences 24 hours of sunlight, and during the winter (*A*), it is totally in the dark. During the equinoxes (*B* & *D*), the Sun's rays are visible at both poles, and all locations on Earth will have 12 hours of sunlight. The constellation Lyra will be visible at night, during the summer months. It will not be visible in the winter because it is up in the sky during the day being outshined by the Sun.

1. Which planet is approximately 20 times farther from the Sun than Earth is?

   (1) Jupiter
   (3) Uranus
   (2) Saturn
   (4) Neptune   1 _____

2. Which planet would float if it could be placed in water?

   (1) Mercury
   (3) Saturn
   (2) Earth
   (4) Jupiter   2 _____

3. Which scale diagram best compares the size of Earth with the size of Venus?

   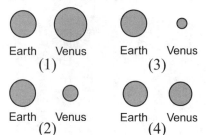

   3 _____

4. Which planet's orbit around the Sun is most nearly circular?

   (1) Mercury
   (3) Earth
   (2) Neptune
   (4) Venus   4 _____

5. Terrestrial planets move more rapidly in their orbits than the Jovian planets because terrestrial planets are

   (1) rotating on a tilted axis
   (2) more dense
   (3) more massive
   (4) closer to the Sun   5 _____

6. The Moon revolves around the center of

   (1) Polaris
   (3) the Milky Way Galaxy
   (2) Aldebaran
   (4) Earth   6 _____

7. How do Jupiter's density and period of rotation compare to Earth's?

   (1) Jupiter is less dense and has a longer period of rotation.
   (2) Jupiter is less dense and has a shorter period of rotation.
   (3) Jupiter is more dense and has a longer period of rotation.
   (4) Jupiter is more dense and has a shorter period of rotation.   7 _____

8. Which graph best represents the relationship between a planet's average distance from the Sun and the time the planet takes to revolve around the Sun?

   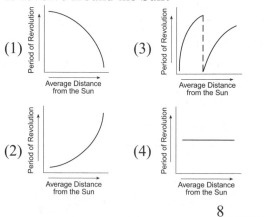

   8 _____

9. Which planet takes more time to complete one rotation on its axis than to complete one revolution around the Sun?

   (1) Mercury
   (3) Mars
   (2) Venus
   (4) Jupiter   9 _____

10. The Sun's position in space is best described as the approximate center of

    (1) a constellation
    (2) the universe
    (3) the Milky Way galaxy
    (4) our solar system   10 _____

11. Which of the following planets has the lowest average density?

    (1) Mercury   (3) Earth
    (2) Venus   (4) Mars   11 _____

12. Which statement correctly compares the size, composition, and density of Neptune to Earth?

    (1) Neptune is smaller, more gaseous, and less dense.
    (2) Neptune is larger, more gaseous, and less dense.
    (3) Neptune is smaller, more solid, and more dense.
    (4) Neptune is larger, more solid, and more dense.   12 _____

13. Compared to Mars, Mercury moves more rapidly in its orbit because Mercury

    (1) is larger
    (2) is more dense
    (3) is closer to the Sun
    (4) has a more elliptical orbit   13 _____

14. Which planet is approximately thirty times farther from the Sun than Earth is?

    (1) Jupiter   (3) Uranus
    (2) Saturn   (4) Neptune   14 _____

15. Which celestial feature is largest in actual size?

    (1) the Moon   (3) the Sun
    (2) Jupiter   (4) the Milky Way galaxy   15 _____

16. The planets known as "gas giants" include Jupiter, Uranus, and

    (1) Venus   (3) Mars
    (2) Saturn   (4) Earth   16 _____

17. Many meteors are believed to be fragments of celestial objects normally found between the orbits of Mars and Jupiter. These objects are classified as

    (1) stars   (3) planets
    (2) asteroids   (4) moons   17 _____

18. The diagram below represents two planets in our solar system drawn to scale, Jupiter and planet A.

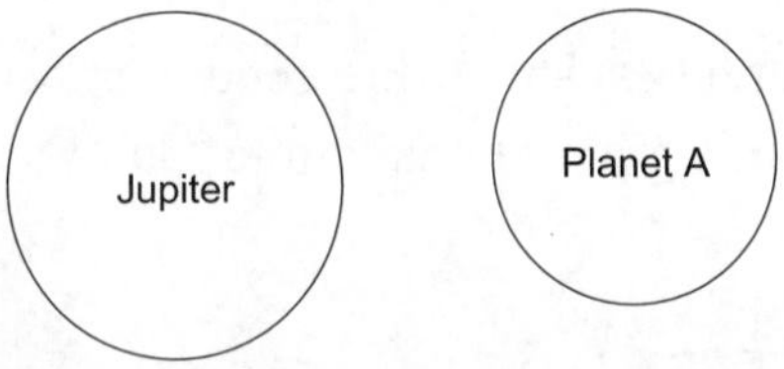

    Planet A most likely represents

    (1) Earth   (3) Saturn
    (2) Venus   (4) Uranus   18 _____

19. Why are some constellations visible to New York State observers at midnight during April, but not visible at midnight during October?

    (1) Constellations move within our galaxy.
    (2) Constellations have elliptical orbits.
    (3) Earth revolves around the Sun.
    (4) Earth rotates on its axis.   19 _____

Note: Question 20 has only three choices.

20. Compared to the average density of the terrestrial planets (Mercury, Venus, Earth, and Mars), the average density of the Jovian planets (Jupiter, Saturn, Uranus, and Neptune) is

    (1) less
    (2) greater
    (3) the same   20 _____

21. Which graph best represents the relative periods of rotation of Mercury, Venus, Earth, and Mars?

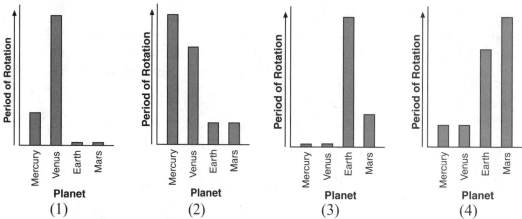

21_____

22. Which graph best shows the general relationship between a planet's distance from the Sun and the Sun's gravitational attraction to the planet?

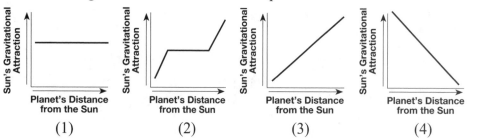

22_____

23. The same side of the Moon always faces Earth because the

(1) Moon's period of rotation is longer than its period of revolution around Earth
(2) Moon's period of rotation is shorter than its period of revolution around Earth
(3) Moon rotates once as it completes one revolution around Earth
(4) Moon does not rotate as it completes one revolution around Earth

23_____

Base your answers to question 24 on the data table, which provides information about four of Jupiter's moons.

24. *a*) Identify the planet in our solar system that is closest in diameter to Callisto.

_____

Data Table

Moons of Jupiter	Density (g/cm³)	Diameter (km)	Distance from Jupiter (km)
Io	3.5	3630	421,600
Europa	3.0	3138	670,900
Ganymede	1.9	5262	1,070,000
Callisto	1.9	4800	1,883,000

b) Which moon of Jupiter is almost the same size of Earth's moon? _____

c) Which moon of Jupiter has the longest period of revolution? _____

Solar System Data

Base your answers to question 25 on the accompanying diagram.

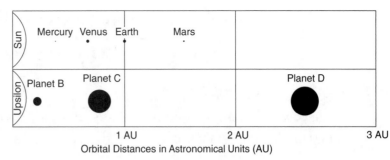

Orbital Distances in Astronomical Units (AU)

25. a) Planet *D*'s diameter is 10 times greater than Earth's diameter.

What planet in our solar system has a diameter closest in size to the diameter of planet *D*? _____

b) Why does Planet *B* revolve faster than Planet *C*? _____

Base your answers to question 26 on the diagram.

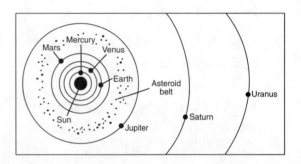

26. a) What is the average distance, in millions of kilometers, from the Sun to the asteroid belt?

_____ mkm

b) What planets revolve faster than our Earth? _____

27. a) Using the data table, on the graph below, draw a line to indicate the general relationship between a planet's average distance from the Sun and its average orbital velocity.

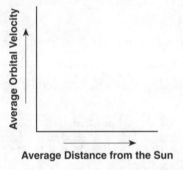

Data Table

Planet	Average Distance from Sun (millions of km)	Average Surface Temperature (°C)	Average Orbital Velocity (km/sec)
Mercury	58	167	47.9
Venus	108	457	35.0
Earth	150	14	29.8
Mars	228	−55	24.1
Jupiter	778	−153	13.1
Saturn	1427	−185	9.7
Uranus	2869	−214	6.8
Neptune	4496	−225	5.4

b) What causes Venus to have a higher average surface temperature compared to Mercury, which is closer to the Sun than Venus? _____

28. Explain why an observer in New York State sees some different constellations in the night sky during the winter months than the summer months.

_____

29. State the shortest number of days it takes an Earth observer to see the same phase of the Moon twice.

_____

1. 3   Open to the Solar System Data chart. Using the Mean Distance from Sun column, the Earth's distance is given as 149.6 million km or close to 150 million km. Multiplying this by 20, we get 3,000 million km. Uranus' distance is given as 2,871 million km, which is close to 3,000 million km.

2. 3   Any object with a density less than 1 g/cm^3 (that of water) will float in water. Saturn has the lowest density of all planets, having a density value of 0.7 g/cm^3 as shown in the Solar System Data chart.

3. 4   In the Solar System Data chart, locate the Equatorial Diameter column. Notice that the diameters of Earth and Venus are very close in size, with Earth being slightly larger in diameter.

4. 4   The Eccentricity of Orbit column gives the eccentricity for each planet. The lower the eccentricity (e) value is, the less elliptical (more circular) the orbit will be. Venus has the lowest e value of 0.007. This makes Venus orbit almost round, but it is still slightly elliptical.

5. 4   Terrestrial planets include Mercury, Venus, Earth and Mars. These planets revolve faster around the Sun than the Jovian planets (Jupiter, Saturn, Uranus and Neptune) because they are closer to the Sun and thus have a stronger gravitational attraction to the Sun.

6. 4   The Moon's orbit is controlled by the gravitational attraction of the Earth, causing the Moon to revolve around our planet.

7. 2   Open to the Solar System Data chart and locate the Density and Period of Rotation columns. Jupiter is a giant gaseous planet, having a density value (1.3 g/cm^3) less than the Earth (5.5 g/cm^3). Jupiter also has the fastest period of rotation of any planet, having a day just under 10 hours.

8. 2   As shown in the Period of Revolution column, as planets' distances from the Sun increase, their period of revolution increases. This is caused by the decreasing gravitational attraction between the planets and the Sun as their distance from the Sun increases.

9. 2   The Period of Rotation column shows Venus' rotation, being its day, as 243 d. The Period of Revolution, being its year, is 224.7 d. This makes its day longer than its year.

10. 4   The Sun is the center of our solar system. Its gravity controls the motion of all objects within our solar system.

Properties of Common Minerals

LUSTER	HARD-NESS	CLEAVAGE	FRACTURE	COMMON COLORS	DISTINGUISHING CHARACTERISTICS	USE(S)	COMPOSITION*	MINERAL NAME
Metallic luster	1–2	✔		silver to gray	black streak, greasy feel	pencil lead, lubricants	C	Graphite
Metallic luster	2.5	✔		metallic silver	gray-black streak, cubic cleavage, density = 7.6 g/cm³	ore of lead, batteries	PbS	Galena
Metallic luster	5.5–6.5		✔	black to silver	black streak, magnetic	ore of iron, steel	Fe_3O_4	Magnetite
Metallic luster	6.5		✔	brassy yellow	green-black streak, (fool's gold)	ore of sulfur	FeS_2	Pyrite
Either	5.5 – 6.5 or 1		✔	metallic silver or earthy red	red-brown streak	ore of iron, jewelry	Fe_2O_3	Hematite
Nonmetallic luster	1	✔		white to green	greasy feel	ceramics, paper	$Mg_3Si_4O_{10}(OH)_2$	Talc
Nonmetallic luster	2		✔	yellow to amber	white-yellow streak	sulfuric acid	S	Sulfur
Nonmetallic luster	2	✔		white to pink or gray	easily scratched by fingernail	plaster of paris, drywall	$CaSO_4 \cdot 2H_2O$	Selenite gypsum
Nonmetallic luster	2–2.5	✔		colorless to yellow	flexible in thin sheets	paint, roofing	$KAl_3Si_3O_{10}(OH)_2$	Muscovite mica
Nonmetallic luster	2.5	✔		colorless to white	cubic cleavage, salty taste	food additive, melts ice	NaCl	Halite
Nonmetallic luster	2.5–3	✔		black to dark brown	flexible in thin sheets	construction materials	$K(Mg,Fe)_3$ $AlSi_3O_{10}(OH)_2$	Biotite mica
Nonmetallic luster	3	✔		colorless or variable	bubbles with acid, rhombohedral cleavage	cement, lime	$CaCO_3$	Calcite
Nonmetallic luster	3.5	✔		colorless or variable	bubbles with acid when powdered	building stones	$CaMg(CO_3)_2$	Dolomite
Nonmetallic luster	4	✔		colorless or variable	cleaves in 4 directions	hydrofluoric acid	CaF_2	Fluorite
Nonmetallic luster	5–6	✔		black to dark green	cleaves in 2 directions at 90°	mineral collections, jewelry	$(Ca,Na)(Mg,Fe,Al)$ $(Si,Al)_2O_6$	Pyroxene (commonly augite)
Nonmetallic luster	5.5	✔		black to dark green	cleaves at 56° and 124°	mineral collections, jewelry	$CaNa(Mg,Fe)_4(Al,Fe,Ti)_3$ $Si_6O_{22}(O,OH)_2$	Amphibole (commonly hornblende)
Nonmetallic luster	6	✔		white to pink	cleaves in 2 directions at 90°	ceramics, glass	$KAlSi_3O_8$	Potassium feldspar (commonly orthoclase)
Nonmetallic luster	6	✔		white to gray	cleaves in 2 directions, striations visible	ceramics, glass	$(Na,Ca)AlSi_3O_8$	Plagioclase feldspar
Nonmetallic luster	6.5		✔	green to gray or brown	commonly light green and granular	furnace bricks, jewelry	$(Fe,Mg)_2SiO_4$	Olivine
Nonmetallic luster	7		✔	colorless or variable	glassy luster, may form hexagonal crystals	glass, jewelry, electronics	SiO_2	Quartz
Nonmetallic luster	6.5–7.5		✔	dark red to green	often seen as red glassy grains in NYS metamorphic rocks	jewelry (NYS gem), abrasives	$Fe_3Al_2Si_3O_{12}$	Garnet

*Chemical symbols:

Al = aluminum	Cl = chlorine	H = hydrogen	Na = sodium	S = sulfur
C = carbon	F = fluorine	K = potassium	O = oxygen	Si = silicon
Ca = calcium	Fe = iron	Mg = magnesium	Pb = lead	Ti = titanium

✔ = dominant form of breakage

Overview:

Imagine if we didn't have minerals: say goodbye to your cell phone, your high definition TV, and to your future car. Even you would cease to exist without those precious dissolved minerals that are so vital for the proper functioning of your body. Our world depends on these solid, inorganic substances. Minerals can be an element, like gold (Au) or copper (Cu), or a simple compound like halite (NaCl); they can also be a more complex compound, as in the mineral talc ($Mg_3Si_4O_{10}(OH)_2$). Mineral properties are controlled by the internal arrangement of the atoms, or simply, how the atoms are bonded together. For example, the composition of diamond and graphite is the same, having only carbon (C) atoms, but their internal atomic arrangements are different. This difference produces two vastly different minerals. The brilliant diamond is the hardest mineral known, while graphite is a soft, greasy, dull mineral.

When minerals are joined together, the solid mass forms a rock. The rock granite is composed of the minerals potassium feldspar, quartz, biotite, and amphibole in various amounts. Mineralogists have identified more than 3,000 minerals, but less than 100 minerals are common and only about 10 elements make up most minerals. So, how do we identify minerals? By testing their physical and chemical properties. These tests might include the mineral's appearance; the crystal shape, luster, hardness, streak, acid test, cleavage or fracture; its form of breakage; and at times how the mineral feels or smells. These properties and others, along with the mineral composition, can be found in the Properties of Common Minerals chart.

The Chart:

The 21 given minerals are classified first by their luster – the way they reflect light. This classification, Metallic or Non-Metallic, is given on the far left of this chart. The next identifying property is hardness, based on a scale of 1 to 10. A mineral can scratch any mineral with a hardness that is less than its own rating. The next column indicates whether the mineral shows cleavage or fracture. If the pattern of breakage of a mineral shows at least one smooth face (side), it has cleavage. The opposite of cleavage is fracture, in which an irregular surface breakage pattern occurs. Common color is given, but one needs to be careful because many minerals show more than one color due to different impurities within the mineral. The Distinguishing Characteristics column gives additional information about the mineral. In this column, streak is mentioned for some of the minerals. Streak is the color of the mineral's powder when that mineral is rubbed on an unglazed porcelain tile. The streak color may not be the same as the color of the mineral. The Use(s) column lists the particular uses for the mineral. The Composition column shows the element or chemical formula of the mineral. At the bottom of this chart, the chemical symbols of elements are given. In the Mineral Name column, locate Potassium feldspar; it also goes by the name orthoclase. Two other minerals are shown to be identified by other names, amphibole and pyroxene.

Additional Information:

- Four minerals on this chart are considered to be ores. Ore deposits contain valuable substances that are mined for economic purposes. Most metals are found in ores.

- Limestone and marble will react by bubble with hydrochloric acid because both rocks have the mineral calcite in their composition.

- Minerals are non-renewable and need to be recycled.

Diagrams:

1. **Streak** – The diagram shows a method for determining the physical property of a mineral known as streak. Streak is the color of the powder of the mineral when it is rubbed on porcelain tile. Some minerals exhibit different streak color compared to their common color. For example, the mineral galena is steel-gray in color, but its streak is black, and calcite is often colorless, but its streak is white.

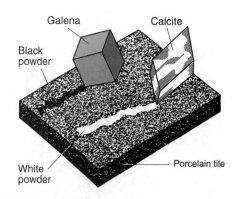

2. **Hardness** – Hardness of a mineral plays an important role in identifying a mineral. The diagram shows a penny scratching the surface of the mineral calcite. Calcite's hardness number is 3, while a penny's hardness number is 3.5. Calcite can be scratched by any mineral with a hardness higher than 3, but it can scratch any mineral with a hardness lower than 3.

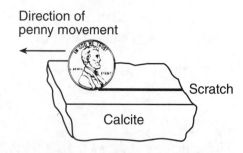

3. **Acid Test** – Calcite ($CaCO_3$), the main mineral in limestone and marble, chemically reacts to acidic solutions. A drop of hydrochloric acid (HCl) will cause these rocks to bubble, releasing carbon dioxide (CO_2). The mineral dolomite, when powdered, will also react to an acid. Acidic rain slowly dissolves calcite within limestone strata, forming caverns.

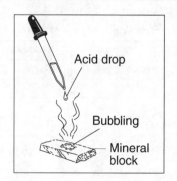

4. **Cleavage** – Cleavage is the way certain minerals split into smooth surfaces when under stress or struck. The mineral mica cleaves into flat, thin sheets, while halite exhibits a cubic cleavage and, when struck, forms smaller cubes. Cleavage is due to the mineral's regular internal arrangement of atoms. Fracture is the opposite of cleavage, having an irregular breakage pattern. The minerals quartz and garnet will fracture when struck.

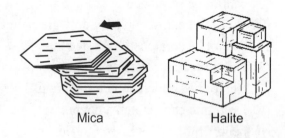

5. **Crystal Shape** – The diagram shows the crystal shapes of two minerals, quartz and halite. Both exhibit different crystal shapes because of the internal arrangement of the atoms. Crystal shape is a defining characteristic property of a mineral.

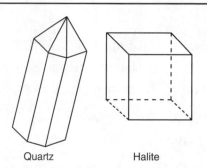

1. Part of a gemstone's value is based on the way a gemstone shines in reflected light. The way a mineral reflects light is described as the mineral's

   (1) fracture   (3) luster
   (2) hardness   (4) streak   1 _____

2. Which mineral will scratch glass (hardness = 5.5), but not pyrite?

   (1) gypsum   (3) orthoclase
   (2) fluorite   (4) quartz   2 _____

3. The table below shows the hardness of four common materials.

   Hardness of Four Materials

Material	Hardness
human fingernail	2.5
copper penny	3.3
window glass	4.5
steel nail	6.5

   Which statement best describes the hardness of the mineral dolomite?

   (1) Dolomite can scratch window glass, but cannot be scratched by a fingernail.
   (2) Dolomite can scratch window glass, but cannot be scratched by a steel nail.
   (3) Dolomite can scratch a copper penny, but cannot be scratched by a fingernail.
   (4) Dolomite can scratch a copper penny, but cannot be scratched by a steel nail.   3 _____

4. An unidentified mineral that is softer than calcite exhibits a metallic luster and cubic cleavage. This mineral most likely is

   (1) galena   (3) halite
   (2) pyrite   (4) pyroxene   4 _____

5. Which mineral leaves a green-black powder when rubbed against an unglazed porcelain plate?

   (1) galena   (3) hematite
   (2) graphite   (4) pyrite   5 _____

6. Which mineral scratches dolomite and is scratched by olivine?

   (1) galena   (3) potassium feldspar
   (2) quartz   (4) muscovite mica   6 _____

7. Which statement about the minerals plagioclase feldspar, selenite gypsum, biotite mica, and talc can best be inferred from the Mineral chart?

   (1) These minerals have the same chemical and physical properties.
   (2) These minerals have different chemical properties, but they have similar physical properties.
   (3) These minerals have different physical and chemical properties, but they have identical uses.
   (4) The physical and chemical properties of these minerals determine how humans use them.   7 _____

8. Minerals from this chart are found in several different rocks. Which two rocks are primarily composed of a mineral that bubbles with acid?

   (1) limestone and marble
   (2) granite and dolostone
   (3) sandstone and quartzite
   (4) slate and conglomerate   8 _____

9. Which home-building material is made mostly from the mineral gypsum?

(1) plastic pipes
(2) window glass
(3) drywall panels
(4) iron nails   9 _____

10. The internal atomic structure of a mineral most likely determines the mineral's

(1) hardness, cleavage, and crystal shape
(2) origin, exposure, and fracture
(3) size, location, and luster
(4) color, streak, and age   10 _____

11. The accompanying table shows some observed physical properties of a mineral. Based on these observations, the elements that make up this mineral's composition are

(1) sulfur and lead
(2) sulfur, oxygen, and hydrogen
(3) oxygen, silicon, hydrogen, and magnesium
(4) oxygen, silicon, aluminum, and iron   11 _____

Physical Property	Observation
color	white
hardness	scratched by the mineral calcite
distinguishing characteristic	feels greasy
cleavage/fracture	shows some definite flat surfaces

Base your answers to question 12 on the table below and on your knowledge of Earth science. The table shows the elements and their percent compositions by mass in the five minerals present in a rock sample.

Elements and Their Compositions by Mass in Five Minerals

Minerals Present in Rock Sample	Element (percent by mass)									
	Al	Ca	Fe	H	K	Mg	Na	O	Si	Ti
Amphibole	6.2	3.0	29.7	0.2	–	3.7	1.8	31.7	12.8	10.9
Plagioclase feldspar	9.7	–	–	–	14.2	–	–	46.3	29.8	–
Garnet	10.9	–	33.8	–	–	–	–	38.7	16.6	–
Muscovite mica	20.3	–	–	0.5	9.8	–	–	48.2	21.2	–
Quartz	–	–	–	–	–	–	–	53.2	46.8	–

12. *a*) Identify *one* use for the mineral garnet. _____

b) Identify *one* mineral in this rock sample that can scratch the mineral olivine. _____

c) All five of the minerals listed in the table are silicate minerals because they contain the elements silicon and oxygen. State the name of *one* other mineral found on the "Properties of Common Minerals" chart that is a silicate mineral.

_____

13. Which two minerals have cleavage planes at right angles?

    (1) biotite mica and muscovite mica
    (2) sulfur and amphibole
    (3) quartz and calcite
    (4) potassium feldspar and pyroxene

    13 _____

14. How are the minerals biotite mica and muscovite mica different?

    (1) Biotite mica is colorless, but muscovite mica is not.
    (2) Biotite mica contains iron and/or magnesium, but muscovite mica does not.
    (3) Muscovite mica scratches quartz, but biotite mica does not.
    (4) Muscovite mica cleaves into thin sheets, but biotite mica does not.

    14 _____

15. The photograph shows a piece of halite that has been recently broken. Which physical property of halite is demonstrated by this pattern of breakage?

    (1) hardness   (3) cleavage
    (2) streak   (4) luster

    15 _____

16. Which mineral is mined for its iron content?

    (1) hematite   (3) galena
    (2) fluorite   (4) talc

    16 _____

17. Which rock would be the best source of the mineral garnet?

    (1) basalt   (3) schist
    (2) limestone   (4) slate

    17 _____

18. Which mineral is commonly mined as a source of the element lead (Pb)?

    (1) galena   (3) magnetite
    (2) quartz   (4) gypsum   18 _____

19. The minerals talc, muscovite mica, quartz, and olivine are similar because they

    (1) have the same hardness
    (2) are the same color
    (3) contain silicon and oxygen
    (4) break along cleavage planes   19 _____

20. The mineral graphite is often used as

    (1) a lubricant
    (2) an abrasive
    (3) a source of iron
    (4) a cementing material   20 _____

21. Quartz and halite have different crystal shapes primarily because

    (1) light reflects from crystal surfaces
    (2) energy is released during crystallization
    (3) of impurities that produce surface variations
    (4) of the internal arrangement of the atoms   21 _____

22. Which minerals contain the two most abundant elements by mass in Earth's crust?

    (1) fluorite and calcite
    (2) magnetite and pyrite
    (3) amphibole and quartz
    (4) galena and sulfur   22 _____

23. Mohs mineral hardness scale and the chart showing the approximate hardness of some common objects. Which statement is best supported by this scale?

(1) A fingernail will scratch calcite, but not quartz.
(2) A fingernail will scratch quartz, but not calcite.
(3) A piece of glass can be scratched by quartz, but not by calcite.
(4) A piece of glass can be scratched by calcite, but not by quartz.

Moh's Mineral Hardness Scale	
Talc	1
Gypsum	2
Calcite	3
Fluortie	4
Apatite	5
Feldspar	6
Quartz	7
Topaz	8
Corundum	9
Diamond	10

Approximate Hardness of Common Objects

Fingernail (2.5)
Copper penny (3.5)
Iron nail (4.5)
Glass (5.5)
Steel file (6.5)
Streak plate (7.0)

22 _____

The diagram shows three minerals with three different physical tests, *A*, *B*, and *C*, being performed on them.

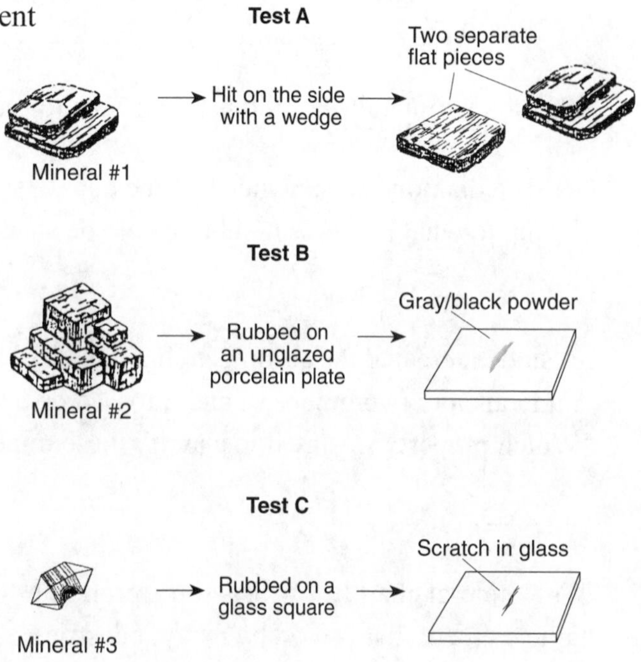

Test A

Mineral #1 → Hit on the side with a wedge → Two separate flat pieces

24. *a)* Which sequence correctly matches each test, *A*, *B*, and *C*, with the mineral property tested?

(1) *A*—cleavage; *B*—streak; *C*—hardness
(2) *A*—cleavage; *B*—hardness; *C*—streak
(3) *A*—streak; *B*—cleavage; *C*—hardness
(4) *A*—streak; *B*—hardness; *C*—cleavage

a _____

Test B

Mineral #2 → Rubbed on an unglazed porcelain plate → Gray/black powder

b) The results of all three physical tests shown are most useful for determining the

(1) rate of weathering of the minerals
(2) identity of the minerals
(3) environment where the minerals formed
(4) geologic period when the minerals formed

Test C

Mineral #3 → Rubbed on a glass square → Scratch in glass

b _____

25. Explain the difference between luster and streak.

_____

26. How can you tell the difference between calcite and halite.

_____

27. What is an ore?

_____

28. What element is present in dolomite that is not present in calcite? _____

29. Of the following minerals: garnet, plagioclase feldspar, pyroxene, and selenite gypsum, which one would most likely become rounded at the fastest rate when tumbled along a stream bottom?

_____

Base your answers to question 30 on the hardness of the minerals talc, quartz, halite, sulfur, and fluorite.

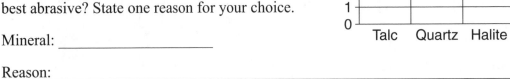

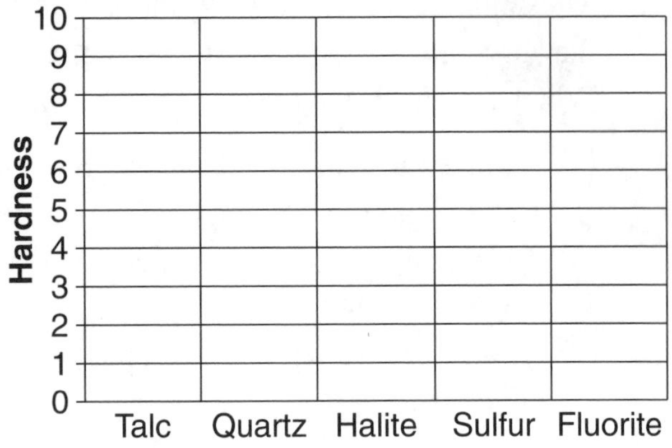

30. *a*) On the grid, construct a bar graph to represent the hardness of these minerals.

    b) Which mineral shown on the grid would be the best abrasive? State one reason for your choice.

    Mineral: _____

    Reason: _____

    c) Which mineral(s) would halite be able to scratch?_____

    d) If a diamond was included on the above bar graph, up to what hardness number would be shade in? _____

31. A student created the accompanying table by classifying six minerals into two groups, *A* and *B*, based on a single property. Which property was used to classify these minerals?

    _____

Group A	Group B
olivine	pyrite
garnet	galena
calcite	graphite

The mineral chart below lists some properties of five minerals that are the major sources of the same metallic element that is used by many industries.

Mineral Chart

Mineral Name	Composition	Density (g/cm³)	Hardness	Streak	Nonmetallic Luster	Common Colors
brucite	$Mg(OH)_2$	2.4	2.5-3	white	glassy to waxy	white
carnallite	$KMgCl_3 \cdot 6H_2O$	1.6	2.5	white	greasy	white
dolomite	$CaMg(CO_3)_2$	2.8	3.5-4	white	glassy to waxy	shades of pink
magnesite	$MgCO_3$	3.1	3.5-4.5	white	glassy	white
olivine	$(Fe,Mg)_2SiO_4$	3.3	6.5	white	glassy	green

32. *a*) Which two minerals have compositions that are most similar to calcite?

    _____   and   _____

    b) Which mineral might scratch the mineral fluorite, but would *not* scratch the mineral amphibole? _____

    c) Which mineral has a different common color from its color in powdered form?

    _____

1. 3   The luster of a mineral is classified as being either metallic or nonmetallic. In both cases, the luster is how light is reflected off the mineral. Choice 4 – streak, is the color of the powder when the mineral is rubbed on an unglazed porcelain tile – known as a streak plate.

2. 3   Locate Orthoclase, which is given under the mineral name of Potassium Feldspar. Its hardness is 6, meaning it can scratch all minerals with a hardness less than 6. This would include glass with a hardness of 5.5. Pyrite, with a hardness of 6.5, would scratch orthoclase.

3. 3   From the Properties of Common Mineral chart, dolomite has a hardness of 3.5. Thus, dolomite can scratch all minerals with a hardness less than 3.5, like copper. Dolomite cannot be scratched by a mineral with a hardness less than 3.5, like a fingernail (2.5).

4. 1   Calcite has a hardness of 3, so the unknown mineral must have a hardness less than 3. Galena has a hardness of 2.5, is metallic in luster and shows cubic cleavage.

5. 4   The color of a mineral's powder is its streak. The streak of pyrite, as shown in the Mineral chart, is green-black.

6. 3   The hardness of dolomite is 3.5 and olivine is 6.5. The unknown mineral's hardness must be between these two hardness values. Potassium feldspar (known as orthoclase) has a hardness of 6. It would scratch dolomite, but be scratched by the harder mineral olivine.

7. 4   Every mineral has its own unique physical and chemical properties as determined by its internal atomic arrangement. The properties that a mineral exhibits determine how we use it and benefit from it.

8. 1   The Mineral chart shows that calcite will bubble with acid. In the composition section of limestone (see page 91) and marble (see page 99), both of these rocks contain calcite.

9. 3   Open to the Properties of Common Minerals chart and locate selenite gypsum. Under Use(s) it states "plaster of paris, drywall." Sheetrock (drywall), which makes up most home walls, is primarily the mineral gypsum. Did you ever use it as chalk on sidewalks?

10. 1   Hardness, cleavage and crystal shape are excellent properties that are used to identify an unknown mineral. The mineral's internal atomic arrangement controls these properties. The age, size, mass and origin of a mineral are not identifying properties of a mineral.

11. 3   The observable properties given in the chart match up with the properties of the mineral talc. The composition of talc is $Mg_3Si_4O_{10}(OH)_2$. Using the chemical symbols found at the bottom of this chart, the elements that make up talc match the ones given in choice 3.

12. *a*)   Answer: Jewelry *or* abrasives   Explanation: This is stated in the Mineral chart.

   b)   Answer: Quartz *or* Garnet

   Explanation: Olivine has a hardness of 6.5. Quartz and garnet have a higher hardness value.

   c)   Answer: Olivine *or* Pyroxene *or* Biotite mica *or* Talc *or* Potassium feldspar

   Explanation: As shown in the Composition column, these minerals have Si and O in their chemical formula.

Overview:

A coordinate system of lines known as latitude and longitude is used to locate any position on the Earth. Latitude lines are measured in degrees north or south from the equator. Longitude lines are measured in degrees east or west from the Prime Meridian. Each degree of latitude or longitude is subdivided into smaller units of minutes (') and seconds ("). Exact location of any position on Earth can be established once these coordinates are known. Latitude and longitude numbers are found on three maps within the reference tables.

Latitude:

Latitude lines are parallel to the equator and are frequently referred to as parallels. The longest of all latitude lines is the equator. It separates the Northern and Southern Hemispheres. At the poles, the latitude line is just an imaginary point. Degrees of latitude range from 0° at the equator to 90° at the poles. The uppermost northern border of New York State is halfway between the equator and the North Pole. Thus its latitude is 45° N (see the Generalized Bedrock Geology of NYS map – above Massena, page 52).

Latitude and the Angle of Polaris – Polaris, the North Star, is positioned almost directly over the North Pole. Due to this location, the latitude of any position in the Northern Hemisphere can be found by measuring the altitude (height) of Polaris above the northern horizon. An observer positioned at the North Pole would therefore view Polaris directly overhead, being 90° above the horizon, which is the latitude of the North Pole. At the equator, Polaris would be positioned on the horizon, having an altitude of 0°, which is the latitude of the equator. From these examples, it can be seen that, in the Northern Hemisphere, as one travels northward, the angle of Polaris would increase to a maximum of 90° at the North Pole. If one were to travel southward, the angle of Polaris would decrease to 0° at the equator. As one travels from the Northern Hemisphere to the Southern Hemisphere (crossing the equator), Polaris would disappear as it dips below the horizon.

Remember:

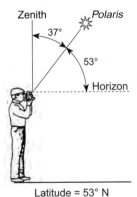

Latitude = 53° N

- The angle of Polaris, measured up from the northern horizon, is equal to the observer's latitude.
  or The latitude of an observer is equal to the angle of Polaris measured up from the northern horizon.

- The height of Polaris changes as an observer travels along a line of longitude in the Northern Hemisphere. But the height of Polaris remains the same when an observer travels along a line of latitude.

- At the North Pole, the angle of Polaris equals 90°.

- At the equator, the angle of Polaris equals 0°.

Longitude:

Longitude lines extend from pole to pole. The Prime Meridian, which runs through Greenwich, England, is the 0° longitude line. All longitude lines are measured east or west from the Prime Meridian. Halfway around the globe from the Prime Meridian, is the maximum longitude of 180°. Longitude lines become closer as they approach the poles, at which point they converge. All positions on a longitude line will have the same solar time.

An observer's longitude can be determined if the time at the Prime Meridian is known when it is solar noon (when the Sun is at its highest altitude) for the observer. This will be explained in greater detail in the *Longitude and Time Zones* section found below.

Time Zones – The rotational speed of the Earth is 15° per hour. Because of this, the globe has been divided into 24 wedge-shape segments – representing time zones – each being 15° of longitude apart. Each time zone is one-hour different from adjacent time zones. Traveling westward, passing into different time zones, time becomes earlier, while traveling eastward time becomes later. On the mainland of United States there are 4 time zones. If it is 10 a.m. in NYS, being in the Eastern time zone, it would be 9 a.m. in the Central time zone, 8 a.m. in the Mountain time zone, and 7 a.m. in the Pacific time zone.

Longitude and Time Zones – Longitude can be calculated if you know the difference between your time and time at the Prime Meridian (referred to as Greenwich Mean Time – GMT) and knowing that each hour difference equals 15° of longitude. The following examples show how this works.

Example 1: What is your longitude if it is solar noon at your location and it is 7 p.m. at the Prime Meridian?

Answer: 105° W

Explanation: Since you are 7 time zones earlier than Greenwich Mean Time, your longitude must be 105° W (7 × 15° = 105°). Since your time is earlier than GMT, you must be west of the Prime Meridian.

Example 2: If an observer's time is 3 p.m. while it is noon at the Prime Meridian, what is the observer's longitude?

Answer: 45° E

Explanation: There is a 3-hour difference, which equals 45° of separation of longitude (3 × 15° = 45°). Since the observer's time is later than that of Greenwich time, the observer must be to the east of Prime Meridian, making the observer's longitude 45° E.

Minutes and Seconds of Latitude and Longitude – A degree of latitude and longitude can be subdivided into smaller units of minutes and seconds. By using these small units, very accurate positioning is possible. For our purpose, only minutes will be explained. Each degree of latitude and longitude is subdivided into 60 equal sections that are called minutes (′). Therefore, 30′ is the halfway position between any two degrees. The illustration on the next page shows how this works with degrees of latitude.

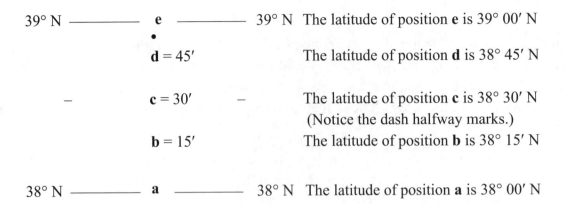

39° N ——————— **e** ——————— 39° N The latitude of position **e** is 39° 00′ N

    •

d = 45′   The latitude of position **d** is 38° 45′ N

—   **c** = 30′   —   The latitude of position **c** is 38° 30′ N
(Notice the dash halfway marks.)

b = 15′   The latitude of position **b** is 38° 15′ N

38° N ——————— **a** ——————— 38° N The latitude of position **a** is 38° 00′ N

This example shows increments of 15′, but remember each degree is divided into 60 minutes. What would you estimate the latitude reading of the dot (•) shown above? A correct answer would be one that is close to a latitude reading of 38° 55′ N. The same system is used with longitude lines. The illustration below shows how this works with degrees of longitude.

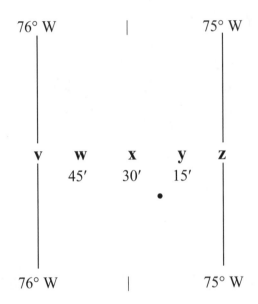

The longitude of position **z** is 75° 00′ W.

The longitude of position **y** is 75° 15′ W.

The longitude of position **x** is 75° 30′ W.
(Notice the dash halfway marks.)

The longitude of position **w** is 75° 45′ W.

The longitude of position **v** is 76° 00′ W.

From the diagram above, what is the estimated longitude position of the dot (•)? _____

A correct answer would be 75° 20′ W (±2°).

Now let's see if you can estimate the full coordinates of a given position.
From the diagram below, what are the coordinates of dot **a** and dot **b**?

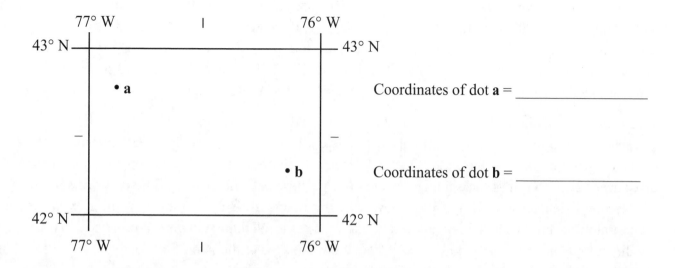

Coordinates of dot **a** = _____

Coordinates of dot **b** = _____

Answer: dot **a** = 42° 45′ N, 76° 50′ W
      (Give yourself credit if your minutes are within ±5′ for both readings.)

Explanation: Latitude readings are always given first. Dot **a** is higher than the halfway position as shown by the – symbol. Knowing that there are 60′ (minutes) in one degree, dot **a** is approximately at the 45′ position, giving an estimated latitude value of 42° 45′ N. For the longitude value, dot **a** is to the left of the 76° 30′ W position (notice the halfway position, represented by the I symbol), being approximately 50′. This gives a longitude reading of 76° 50′ W.

Answer: dot **b** = 42° 20′ N, 76° 10′ W
      (Give yourself credit if your minutes are within ±5′ for both readings.)

Explanation: Dot **b** is below the 42° 30′ N position, being close to 20′ position. This gives a latitude reading of 42° 20′ N. The longitude position of dot **b** is less than the halfway position, (76° 30′ W), being close to the 10′ position. This gives a longitude reading of 76° 10′ W.

Reference Tables Showing Coordinates:

The *Generalized Bedrock Geology of New York State* map, the *Surface Ocean Currents* map and the *Tectonic Plates* map all show coordinate numbers. On the Generalized Bedrock map, the latitude and longitude numbers are shown for New York State. The 30′ reading is shown by the dashed lines halfway between the degrees. To get an accurate reading for the coordinates, use a straight edge to cross the chart, aligning the same degree readings; then estimate the correct minutes value for the given position. For example: What are the coordinates of Syracuse? Solution: Place a straight edge or a ruler on the map such that it passes through the 43° N latitude numbers on both sides of the chart. From this alignment, one can see that Syracuse's latitude is slightly higher than 43° N. A given latitude answer of 43° 05′ N would be a correct estimate. Next, place a straight edge extending through the 76° longitude reading. This shows that a correct longitude estimate would be 76° 10′ W. The coordinates for Syracuse are 43° 05′ N, 76° 10′ W.

On the *Tectonic Plates* map, estimated coordinate readings can be obtained for mantle hot spots and locations of plate boundaries. On this chart, the highest latitude values shown are 70° N and S.

On the *Surface Ocean Currents* map, latitude values are shown on the left side of the map. Although latitude goes up to 90°, the highest latitude reading shown is 80° N and S. Longitude values are given on the top and bottom of this map. The 0° longitude line is the Prime Meridian and on this map it is positioned on the far right. The maximum longitude reading is 180° W or E. The Surface Ocean Currents map has 5 additional lines that you must have an understanding of. (See page 60.)

Tropic of Cancer, 23.5° N – Due to the Earth's tilt and motion of revolution, the Sun's direct rays (perpendicular to the Earth's surface) appear to move a total of 47° (23.5° N to 23.5° S) in six months, – spanning the tropics. When these direct rays reach the farthest northern position, striking the Tropic of Cancer (23.5° N), it is the summer solstice (on or close to June 21) – the first day of summer for the Northern Hemisphere. On this day, as viewed from New York State, the noon Sun has reached its highest altitude, producing the longest duration (length) of insolation (sunlight). For the next six months, the direct rays move south, and as viewed from New York State, the noon Sun continually decreases in altitude causing a decrease in the hours of sunlight.

Tropic of Capricorn, 23.5° S – On the winter solstice (on or close to December 21), the Sun's direct rays have reached its southern most position – hitting the Tropic of Capricorn. On this day, as viewed from New York State, the noon Sun has reached its lowest altitude, producing the shortest duration of insolation. For the next six months, the direct rays move north, and as viewed from New York State, the noon Sun continually increases in altitude causing an increase in the hours of sunlight.

Arctic Circle, 66.5° N – On the summer solstice, all positions on and north of this latitude line would experience 24 hours of sunlight, as the Sun circles above the horizon for 24 hours. On the winter solstice, all positions on and north of the Arctic Circle experience 24 hours of darkness, because the Sun remains below the horizon for 24 hours. Periods of continuous day and night range from one day at the Arctic Circle to six months at the North Pole. On the equinoxes – first day of spring and fall – all positions on the Earth will have 12 hours of sunlight and 12 hours of night. On these days, sunlight hits both the South Pole and the North Pole.

Antarctic Circle, 66.5° S – The events and occurrences at the Antarctic circle are the same as the Arctic Circle (described above), except they take place during the Southern Hemisphere solstices. Remember, for the Southern Hemisphere, the seasons are opposite of our seasons. Periods of continuous day and night range from one day at the Antarctic Circle to six months at the South Pole.

Note: All latitude and longitude questions dealing with these charts will be found in their respected sections.

Additional Information:
- To locate Polaris, the constellation The Big Dipper (Ursa Major) is used. The two "pointer stars," at the end of the bowl, direct a line of sight to the star Polaris.

- Polaris is located at the end of the handle of the Little Dipper (Ursa Minor).

Diagrams:

1. **Coordinates** – Latitude is measured N and S from the equator. The maximum number of degrees for latitude is 90°, located at the poles. Longitude is measured W and E from the Prime Meridian. The maximum number of degrees for longitude is 180°. Notice how all the longitude lines converge at the poles.
The coordinates of position **X** is 50° N, 45° W.

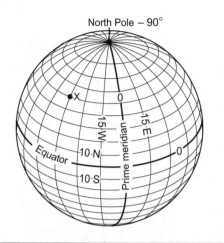

---

2. **Important Latitude Lines** – The diagram shows the Earth in its winter position for the Northern Hemisphere. Notice that the North Pole would rotate in complete darkness, while the South Pole would rotate in complete sunlight. On this date, December 21st, direct sunlight (90°) is located on 23.5° S. Six months later, the Sun's direct rays would be located on 23.5° N. The apparent movement of the Sun's direct rays to new latitude positions is due to the tilt of our axis and the motion of revolution.

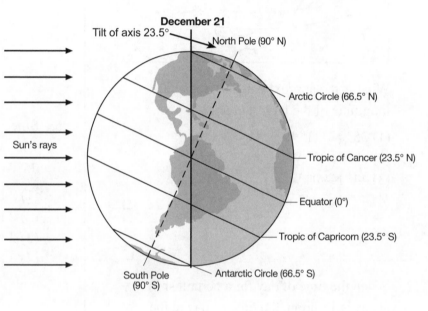

---

3. **Time Zones** – There are 24 time zones each spanning 15° of longitude. This system is based on the rotation speed of the Earth, which is 15°/hour. From any location, if you travel west into a new time zone, the time becomes earlier, while going east the time becomes later. If two cities are on the same longitude line, regardless of the distances, they will have the same solar time.

This polar view diagram represents the equinox where day and night are equal.

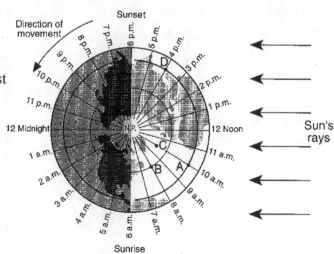

---

1. The map below shows the location and diameter, in kilometers, of four meteorite impact craters, *A*, *B*, *C*, and *D*, found in the United States.

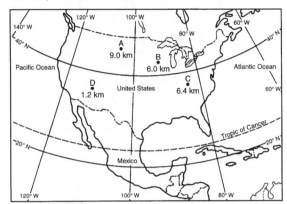

What is the approximate latitude and longitude of the largest crater?

(1) 35° N 111° W
(2) 39° N 83° W
(3) 44° N 90° W
(4) 47° N 104° W   1 _____

2. When the time of day for a certain ship at sea is 12 noon, the time of day at the Prime Meridian (0° longitude) is 5 p.m. What is the ship's longitude?

(1) 45° W   (3) 75° W
(2) 45° E   (4) 75° E   2 _____

3. Of the following choices, the maximum a longitude reading with its correct compass direction is

(1) 90° N   (3) 180° N
(2) 90° E   (4) 180° E   3 _____

4. The model below represents the apparent path of the Sun across the sky on March 21 as seen by an observer on Earth.

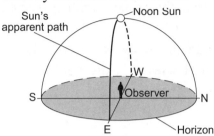

At which latitude is the observer located?

(1) 90° N   (3) 23.5° N
(2) 42° N   (4) 0°   4 _____

5. Since Denver's longitude is 105° W and Utica's longitude is 75° W, sunrise in Denver occurs

(1) 2 hours earlier
(2) 2 hours later
(3) 3 hours earlier
(4) 3 hours later   5 _____

6. The diagram below represents part of Earth's latitude-longitude system.

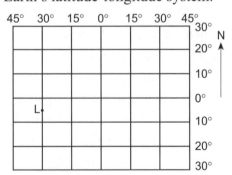

What is the latitude and longitude of point L?

(1) 5° E 30° N   (3) 5° N 30° E
(2) 5° W 30° S   (4) 5° S 30° W   6 _____

7. The diagram below shows the latitude-longitude grid on an Earth model. Points *A* and *B* are locations on the surface.

On Earth, the solar time difference between point *A* and point *B* would be

(1) 1 hour   (3) 12 hours
(2) 5 hours   (4) 24 hours   7 _____

8. *a*) The map below, shows the latitude and longitude of five observers, *A*, *B*, *C*, *D*, and *E*, on Earth.

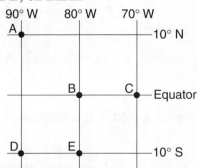

What is the altitude of Polaris (the North Star) above the northern horizon for observer *A*?

(1) 0°   (3) 80°
(2) 10°   (4) 90°   a _____

b) Which two observers would be experiencing the same apparent solar time?

(1) *A* and *C*   (3) *B* and *E*
(2) *B* and *C*   (4) *D* and *E*   b _____

9. The dashed line on the map below shows a ship's route from Long Island, New York, to Florida. As the ship travels south, the star Polaris appears lower in the northern sky each night.

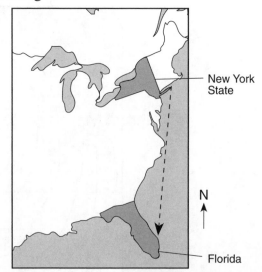

Showing the map of the Eastern part of the United States, the best explanation for this observation is that Polaris

(1) rises and sets at different locations each day
(2) has an elliptical orbit around Earth
(3) is located directly over Earth's Equator
(4) is located directly over Earth's North Pole   9 _____

10. The diagram below shows an observer measuring the altitude of Polaris.

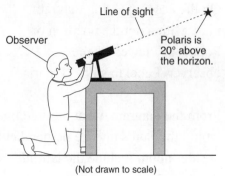

(Not drawn to scale)

What is the latitude of the observer?

(1) 20° N   (3) 70° N
(2) 20° S   (4) 70° S   10 _____

Base your answer to question 11 on the map. The map shows a portion of the Indian Ocean and surrounding landmasses. The location of the epicenter of a large undersea earthquake that occurred on December 26, 2004, is shown by an **X**. The isolines surrounding the epicenter show the approximate location of the first tsunami wave produced by this earthquake in half-hour intervals after the initial earthquake.

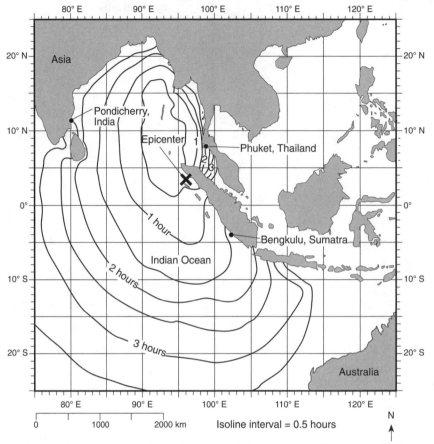

11. *a*) State the latitude and longitude of the epicenter of this earthquake. Include the units and compass directions in your answer. _____

   b) On the latitude readings located on the edge of the map, place a **X** next to the Tropic of Cancer.

12. *a*) On the accompanying diagram, mark with a dot the position of Polaris as viewed by the observer. Label this dot "Polaris."

   b) From the diagram, what latitude line would the Sun's rays be perpendicular at solar noon. Include the latitude direction in your answer.

   _____

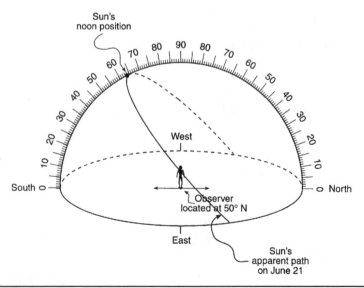

13. Which statement about *Polaris* is best illustrated by the diagrams shown below?

At Equator

At New Orleans, Louisiana

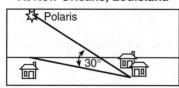

At North Pole

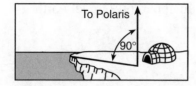

(1) *Polaris* is located in a winter constellation.
(2) *Polaris* is located at the zenith at each location.
(3) *Polaris'* apparent movement through the sky follows a south-to-north orientation.
(4) *Polaris'* altitude is equal to a location's latitude.   13 _____

14. Earth's rate of rotation is approximately

(1) 1° per day
(2) 15° per day
(3) 180° per day
(4) 360° per day   14 _____

15. At which location will the highest altitude of the star *Polaris* be observed?

(1) Equator
(2) Tropic of Cancer
(3) Arctic Circle
(4) central New York State   15 _____

16. The lines on which set of views best represent Earth's latitude system?

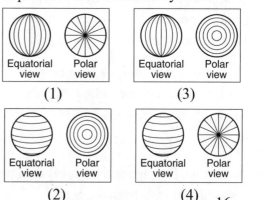

16 _____

17. As a ship crosses the Prime Meridian, an observer on the ship measures the altitude of Polaris at 60°. What is the ship's location?

(1) 60° south latitude and 0° longitude
(2) 60° north latitude and 0° longitude
(3) 0° latitude and 60° east longitude
(4) 0° latitude and 60° west longitude

17 _____

18. What time is it in Greenwich, England (at 0° longitude), when it is noon in Massena, New York having a longitude of 75° W?

(1) 7 a.m.   (3) 5 p.m.
(2) noon   (4) 10 p.m.   18 _____

Note: Question 19 has only three choices.

19. With the aid of a GPS tracking device, a hiker covers 500 miles traveling due west, while maintaining a constant 45° N latitude bearing. During his travel, the angle of Polaris would have

(1) increased
(2) decreased
(3) remain the same   19 _____

The diagram represents four apparent paths of the Sun, labeled *A*, *B*, *C*, and *D*, observed in Jamestown, New York. The June 21 and December 21 sunrise and sunset positions are indicated. Letter *S* identifies the Sun's position on path *C* at a specific time of day. Compass directions are indicated along the horizon.

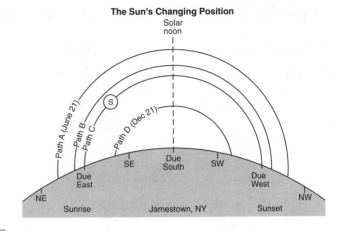

20. *a)* The greatest duration of insolation in Jamestown occurs when the Sun appears to travel along path

(1) *A*   (2) *B*   (3) *C*   (4) *D*   a _____

b) When the Sun appears to travel along path *D* at Jamestown, which latitude on Earth receives the most direct rays from the Sun?

(1) 42° N   (2) 23.5° N   (3) 0°   (4) 23.5° S   b _____

c) How many degrees of latitude does the direct rays of the Sun apparently travel north from Path *D* to being on Path *A*?

(1) 15°   (2) 23.5°   (3) 47°   (4) 66.5°   c _____

21. Based on the Sun's apparent path, where is location *D*?

(1) equator
(2) Tropic of Cancer
(3) Tropic of Capricorn
(4) North Pole

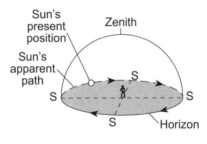

Location D

21 _____

The diagram represents the apparent path of the Sun on the dates indicated for an observer in New York State. The diagram also shows the angle of Polaris above the horizon.

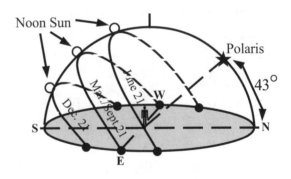

22. *a)* State the latitude of the location represented by the diagram to the nearest degree. Include the latitude direction in your answer.

_____

b) Give a statement on the direction of where the Sun rises and sets on:

December 21: rises _____ sets _____

June 21: rises _____ sets _____

c) At noon, on September 21, give the compass direction of the observers shadow. _____

Base your answer to question 23 on the map.

23. *a)* State the latitude and longitude of points *F* and *G*, to the nearest degree. Include the correct units and compass directions in your answer.

F_____   G_____

b) What is the solar time difference between point 2 and point 3? _____hr

c) Which position would observe sunrise first?
(1) G   (2) 2   (3) 3   (4) F   c_____

d) Give a statement on the observed solar time of position 1 compared to the observed solar time of position 2.

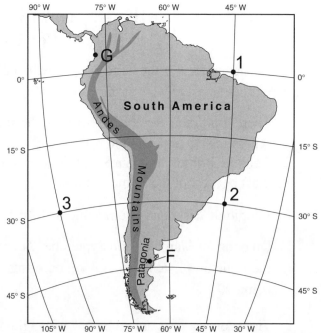

_____

24. State *one* factor, other than the tilt of Earth's axis, that causes seasons to change on Earth. _____

25. At which latitude is *Polaris* observed at an altitude of 66.5°? _____

Base your answers to question 26 on the map, which shows a portion of southwestern United States. On January 17, 1994, an earthquake occurred with an epicenter at Northridge, California.

26. *a)* State the latitude and longitude of Northridge, California. Include the correct units and compass directions in your answer.

_____

b) Oakland is located in the Pacific Time zone. If it is noon in Oakland, what time would it be in Utah, which is located in the Mountain Time zone?_____

27. To an observer in New York State, the hours of daylight increase continuously for six months after the Sun's direct rays strike what degree latitude line? _____°

1. 4   Impact crater *A* is the largest crater with a diameter of 9.0 km. Its position is above the 40° N latitude line and west of the 100° W longitude line. By elimination, only choice 4 could be correct.

2. 3   There is a 5-hour separation between the ship's time and that at the Prime Meridian. Each hour of separation equals one time zone, which is 15 degrees of longitude. The ship must be 75 degrees from the Prime Meridian. Since the ship's time is earlier than that at the Prime Meridian, the ship must be west of the Prime Meridian.

3. 4   Longitude has a maximum value of 180° E and W. Latitude has a maximum value of 90° N and S.

4. 4   On or about March 21 and September 21 are the equinoxes. On this date the Sun's direct rays strike the Equator, 0° latitude. At noon an observer on the equator would see the Sun directly overhead being 90°.

5. 1   Time zones are separated by 15° of longitude. The separation between these two cities is 30°, or 2 hours. Denver is west of Utica, so its time is earlier.

6. 4   Point *L* is south of the equator by about 5 degrees. The longitude position is west of the Prime Meridian line (0° line) by 30 degrees. The coordinate reading would be written as 5° S, 30° W.

7. 2   Position *A* is on the Prime Meridian (0°). Position *B* is on the 75° W longitude line. Time zones are every 15° of longitude. Dividing 75°/15° = 5 hours difference of time.

8. *a)* 2   The altitude of Polaris in the Northern Hemisphere equals the latitude of the observer. Observer *A* is on the 10° N latitude line, thus the altitude of Polaris would be 10° above the northern horizon.

    b) 3   Positions on the same longitude line (in this case, 80° W) would have the same apparent solar time.

9. 4   Polaris is located 90° above the northern axis of the Earth. Due to the curvature of the Earth's surface, as one travels south in the Northern Hemisphere, Polaris will appear lower in the northern sky. At the equator Polaris' altitude is 0°.

10. 1   The angle of Polaris above the northern horizon is equal to the latitude of the observer. The observer must be positioned on the 20° N latitude line in order to see Polaris at 20°.

11. *a)* Answer: 4° N, 96° E (±1°)

    Explanation: Using the coordinate values on the sides of the map, the epicenter **X** has a latitude of 4° N and a longitude of 96° E.

    b) The **X** must be positioned on the 23.5° N latitude line.

12. *a)* Answer: The dot needs to be placed on the globe 50° position up from North.

    Explanation: An observer's latitude will equal the altitude of Polaris up from the northern horizon. From the given diagram, it shows that the observer is located on the 50° N latitude line.

    b) Answer: 23.5° N   Explanation: See page 220, Tropic of Cancer.